Urban Sustainability

The Urban Sustainability Book Series is a valuable resource for sustainability and urban-related education and research. It offers an inter-disciplinary platform covering all four areas of practice, policy, education, research, and their nexus. The publications in this series are related to critical areas of sustainability, urban studies, planning, and urban geography.

This book series aims to put together cutting-edge research findings linked to the overarching field of urban sustainability. The scope and nature of the topic are broad and interdisciplinary and bring together various associated disciplines from sustainable development, environmental sciences, urbanism, etc. With many advanced research findings in the field, there is a need to put together various discussions and contributions on specific sustainability fields, covering a good range of topics on sustainable development, sustainable urbanism, and urban sustainability. Despite the broad range of issues, we note the importance of practical and policy-oriented directions, extending the literature and directions and pathways towards achieving urban sustainability.

The series will appeal to urbanists, geographers, planners, engineers, architects, governmental authorities, policymakers, researchers of all levels, and to all of those interested in a wide-ranging overview of urban sustainability and its associated fields. The series includes monographs and edited volumes, covering a range of topics under the urban sustainability topic, which can also be used for teaching materials.

Javid Ahmad Parray · Nowsheen Shameem ·
A. K. Haghi
Editors

Sustainable Urban Environment and Waste Management

Theory and Practice

Springer

Editors
Javid Ahmad Parray
Environmental Science
GDC Eidgah
Srinagar, Jammu and Kashmir, India

Nowsheen Shameem
Cluster University
Srinagar, Jammu and Kashmir, India

A. K. Haghi
Research Chemistry Center
University of Coimbra
Coimbra, Portugal

ISSN 2731-6483 ISSN 2731-6491 (electronic)
Urban Sustainability
ISBN 978-981-96-1139-3 ISBN 978-981-96-1140-9 (eBook)
https://doi.org/10.1007/978-981-96-1140-9

This Springer imprint is published by the registered company Springer Nature Singapore Pte Ltd.
The registered company address is: 152 Beach Road, #21-01/04 Gateway East, Singapore 189721, Singapore

Contents

Water Pollution Reduction for Sustainable Urban Development

Nurendah Ratri Azhar Rusprayunita, Sepsiana Puspitasari, Hodimatum Mahiroh, Enrika Rahayu Setyani, Veriana Indah Citraningtyas, Windu Syawalina Wahyuningsih, and Annisa Utami Rauf

Abstract The escalating global population growth and rapid urban expansion have heightened the urgency for sustainable practices to elevate water quality standards. This chapter presents a comprehensive overview of the current state of water quality, highlighting alarming levels of pollution that pervade our water sources. From industrial effluents to agricultural runoff, these contaminants pose significant risks to human health and ecosystem stability. A pivotal concern is the management of wastewater originating from diverse human activities, a critical aspect indispensable for sustainability and human welfare. In response, this chapter advocates for a concerted effort to promote responsible water usage through community-led initiatives and educational campaigns. A fundamental strategy involves curtailing groundwater exploitation associated with activities such as hospitality and manufacturing, thereby reducing the risk of communities resorting to contaminated water sources. With the adoption of advanced technologies and sustainable practices for treating industrial and commercial wastewater, substantial reductions in pollutants discharged into water bodies can be achieved. Effective wastewater management necessitates collaborative efforts among governmental bodies, industries, and local communities. Through a holistic approach encompassing education, robust regulatory frameworks, and investments in sustainable technologies, the detrimental impact of human activities on water quality can be mitigated. Prioritizing sustainable wastewater management not only safeguards the health of current and future generations but also paves the path toward a healthier environment and a more resilient future.

Keywords Environmental pollution · Sustainable practices · Water quality · Wastewater management

N. R. A. Rusprayunita · S. Puspitasari · H. Mahiroh · E. R. Setyani · V. I. Citraningtyas · W. S. Wahyuningsih
Department of Biostatistics, Epidemiology and Population Health, Faculty of Medicine, Public Health and Nursing, Universitas Gadjah Mada, Yogyakarta 55281, Indonesia

A. U. Rauf (✉)
Department of Health Behavior, Environment and Social Medicine, Faculty of Medicine, Public Health and Nursing, Universitas Gadjah Mada, Yogyakarta 55281, Indonesia
e-mail: annisaur@ugm.ac.id

J. A. Parray et al. (eds.), *Sustainable Urban Environment and Waste Management*, Urban Sustainability, https://doi.org/10.1007/978-981-96-1140-9_1

1 Introduction

Access to clean water, adequate sanitation facilities, and proper hygiene practices are fundamental determinants of human health. Maintaining our clean and uncontaminated water has become a significant issue as global population densities increase. Inadequate practices associated with industrial and agricultural activities are frequently identified as significant contributors to global water pollution [1]. Global freshwater withdrawals have exhibited a six-fold increase since the nineteenth century. This surge in water consumption, coupled with widespread water pollution from diverse contaminants, presents an emerging challenge: the scarcity of clean water [2]. The earlier concerns are a prominent global risk in the current and forthcoming era. Data compiled by the World Health Organization (WHO) suggests that approximately 2 billion individuals worldwide are deprived of safe drinking water access [3].

Current estimates suggest that a significant portion of industrial and other waste streams, potentially reaching 80%, are discharged directly into the environment without prior treatment. This practice undoubtedly poses a substantial threat to environmental health, particularly impacting water resources. Heavy metals such as cadmium, arsenic, and chromium are the most concerning pollutants frequently identified in these untreated wastewater discharges. The concentration of these toxic substances within the environment is demonstrably increasing with ongoing industrialization [4–6]. The agricultural sector also contributes significantly to water pollution. The improper application and disposal of pesticides and various poultry-related chemical substances also directly jeopardize human health. These contaminants are transported through various environmental pathways, leading to a decline in water quality and a diminished supply of clean water [7]. This condition puts nearly 5 billion people at risk of water stress by 2050 [8, 9].

A significant manifestation of water stress is groundwater depletion, which arises from prolonged and excessive freshwater withdrawal exceeding natural recharge rates. This depletion has many negative consequences, including declining overall water availability, consequently increasing the likelihood of water scarcity possibilities [10–12]. Furthermore, excessive groundwater withdrawal in coastal regions exacerbates the risk of saltwater intrusion. This phenomenon occurs when saltwater resettles into freshwater aquifers due to the pressure differential caused by depletion. Additionally, groundwater depletion can worsen the impact of pre-existing pollutants within the aquifer [13]. As the volume of freshwater diminishes, the concentration of these contaminants increases due to reduced dilution capacity so that the quality of the resource deteriorates, potentially posing health risks for human consumption and hindering crop production in agricultural applications [14, 15].

Ingestion of water contaminated with harmful substances has been demonstrably linked to many adverse health effects in humans [16]. Numerous scientific studies have established that prolonged exposure can lead to bioaccumulation and contribute to the development of chronic diseases, including neurotoxicity, various forms of cancer, cardiovascular disease, tissue necrosis, asthma, reproductive disorders including infertility, diabetes and other metabolic syndrome, neurodegenerative disorders, genetic alteration, and congenital malformations in offspring [17–19].

Given the fundamental role water plays in sustaining human life, the degradation of water bodies is linked to a decline in the quality of life for individuals and communities. Conversely, numerous research has demonstrated that improved access to clean water, intertwined with good sanitation and hygiene practices, leads to better health outcomes across populations [20]. Mitigating water pollution necessitates a multi-pronged approach. On the industrial front, adopting water-efficient technologies holds significant promise to dwindle water consumption within production processes. Furthermore, the development and implementation of clear and concise government policies and regulations on wastewater treatment are crucial for safeguarding the long-term health of our water resources [21, 22]. Lastly, fostering public awareness and encouraging behavioral modifications that promote responsible water use in daily life are fundamental to achieving sustainable water management practices [23, 24]. The keys to successful urban development are tackling water pollution and implementing sustainable water management practices. These measures improve the quality of life for city residents by ensuring access to clean water and fostering a healthy living environment. A collaborative effort across various sectors can make a notable difference in protecting this vital resource [25, 26].

2 Current Condition of Our Water

Modern farming methods, rapid urban growth, and the expansion of industrial operations have led to a rise in water waste production. This condition, in turn, directly impacts human well-being by degrading water quality [27]. The decline in water quality poses a significant threat to ecosystems and human health, with groundwater depletion a major concern [28]. The manufacturing industries frequently engages solid, gaseous, and liquid waste directly into the environment. Approximately 80% of liquid waste globally is discharged untreated into rivers and oceans, significantly contributing to widespread water pollution [5, 29, 30]. A study from specific areas of Ogun State, Nigeria, revealed elevated levels of pH, calcium (Ca), lead (Pb), nickel (Ni), manganese (Mn), iron (Fe), cadmium (Cd), and chromium (Cr) exceeding established thresholds. This indicates the groundwater in these regions is unfit for human consumption without prior purification and treatment [31]. Figure 1 shows the findings from a meta-analysis conducted by [32], highlighting substantial differences in the average concentrations of heavy metals (Fe, Mn, Co, Cu, Ni, Zn, Cd, Cr, As) among different countries.

The mining industry undeniably contributes significantly to socio-economic development. However, it also raises significant environmental concerns, particularly regarding its impact on hydrological systems [33]. The mining industry is characterized by high water consumption throughout its extraction and beneficiation processes.

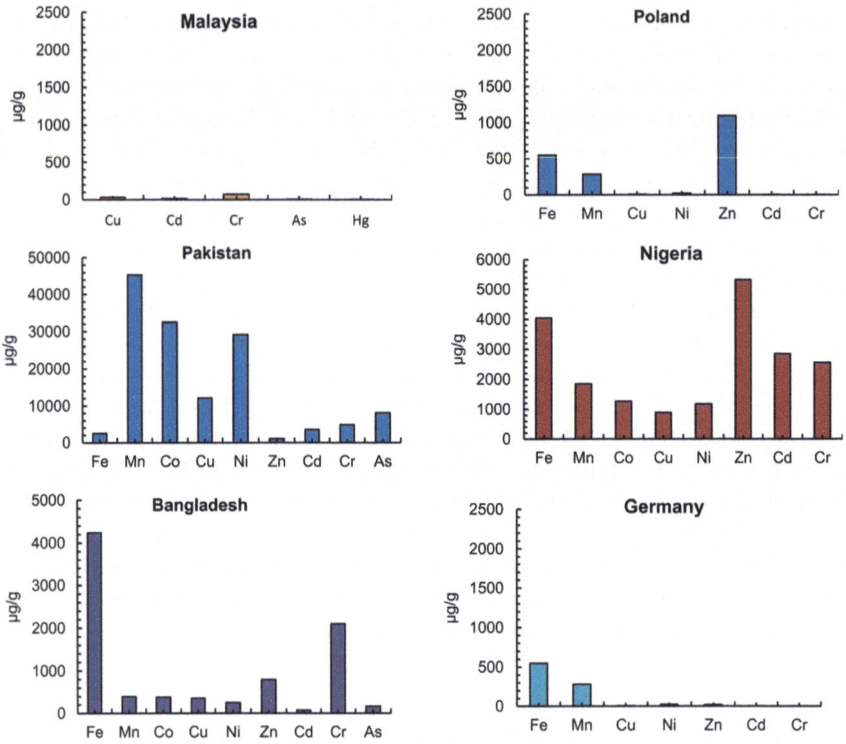

Fig. 1 Mean concentrations of heavy metals in surface water bodies across various countries. Adapted from [32]

This extensive water usage can dissolve and mobilize heavy metal ions, contaminating surrounding soil and potentially generating Acid Mine Drainage (AMD) [34]. AMD characterized by the formation of acidic groundwater with elevated sulfate concentrations and the presence of dissolved heavy metals [35, 36], which worsen the groundwater quality and endanger its surrounding [37]. For instance, the gold mining conducted in the Bétaré-Oya area causes the surrounding water to become acidic, with elemental contamination and high concentrations of total suspended solids (TSS), making it not suitable for public consumption [38].

Like the industrial sector, agricultural activities play a significant role in the deterioration of water quality. The widespread expansion and intensification of irrigation systems, rain-fed agriculture, livestock production, and inland aquaculture lead to increased water consumption and utilization of anthropogenic materials. These materials include agrochemicals, animal feed additives, antibiotics, and pesticides [39]. Applying fertilizers commonly used on crops like soybeans, wheat, rice, and maize can lead to nitrate leaching. This leaching process contaminates groundwater reserves, rendering them unsuitable for human consumption [40]. Such practices exacerbate the depletion of groundwater resources across all geological formations, including both impermeable (aquifuges) and permeable (aquifers) layers [41]. Furthermore, water pollution associated with agricultural practices carries a significant economic burden. This burden extends across various domains, such as human health due to heavy metal contamination in drinking water, which can lead to health problems, ecosystem aspects including toxic algae blooms and salinization, and some recreational or cultural activities. Based on the evidence mentioned above, the urgent need for stricter regulations on agricultural practices to safeguard water quality becomes imperative.

Accelerated population growth fosters an increase in population density, particularly within urban centers compared to rural areas. This human population expansion fuels economic development, often characterized by rapid industrialization. However, these industrial activities necessitate significant water resources [42]. Urbanization generates water pollution through anthropogenic sources. These include municipal waste, such as effluent from domestic and commercial activities directly discharged into waterways, and inadequate sewage systems that lead to untreated wastewater [43, 44]. Additionally, ineffective urban planning, including poorly designed infrastructure, often fails to manage stormwater runoff effectively. Engineering-related activities, such as erosion and sedimentation from construction sites, also significantly impact water pollution. These activities can alter waterways, mainly due to impervious surfaces like pavements, buildings, and roads [45]. Furthermore, urban agricultural practices can substantially degrade water quality, worsen the pollution burden.

Beyond anthropogenic sources of water contamination, geological features inherent to our planet can also significantly degrade water quality [46]. Notably, the natural process of mineral leaching plays a crucial role. Arsenic and fluoride are two particularly concerning minerals commonly found in groundwater that are prone to leaching into the surrounding water table [47]. Additionally, erosion and sedimentation are recognized geogenic contributors to water pollution. Sedimentation is driven by a complex interplay of three primary factors: tectonic activity, climatic patterns, and sea level fluctuations [48, 49]. This process has several environmental consequences such as heavy metal infiltration, since erosion can transport heavy metals bound to soil particles and absorbed into water bodies; adsorbed nutrient transport due to sediment can act as carriers for adsorbed nutrients such as phosphorus that can contribute to toxic algae bloom; and pathogen dislodge from land to water bodies [50].

3 Advocating Responsible Water Usage

Currently, the global issue of overconsumption of water is widespread which induces freshwater scarcity [11, 15]. Moreover, human activity like agriculture practices, industry, and urban waste can exacerbate water quality, which is called the degradation of water [37, 39, 51]. World Health Organization (WHO) has recommended a water use minimum of 50 L per person per day. Water use restrictions are also implemented to manage water scarcity such as the case of drought in South Africa which requires residents to reduce water use to less than 50 L per person per day [52]. Water scarcity is also associated with power, poverty, and inequity. Hence, it is very crucial to ensure the effort of water resource management and services in the long-term prospect [53]. According to The Water Resources Group, if water management is carried out as usual, by 2030 there will be a 40% gap between freshwater supply and demand [52].

One of the breakthroughs regarding water scarcity is community-managed water supply (CMWS) [54]. In this context, society is being responsible for the management and operation. The procurement of facilities to support that system including technical provision, and political and institutional assistance, is delegated to external agents [55–57]. Nonetheless, this resolution has drawbacks related to the community level, such as insufficient capacity and systems. Therefore, it requires commitment from community and post-construction assistance, so it can be sustainable [58]. In addition, there is also a community-based water governance (CWG) policy to improve water security through water governance in urban areas. Most case studies showed that CWG has an impact on increasing water quantity, supply reliability, prices, community empowerment, employment, water revenue reduction, and financial viability. The implementation of CWG greatly requires community partnerships with non-governmental organizations, the private sector, public utilities, and community autonomy for the sustainability of the program. However, in reality, the implementation of CWG is still fragmented due to weak capacity and infrastructure, geographical coverage, funding issues, and political interference [59].

Besides water supply management, behavioural water usage also can be a solution to overcome water scarcity. Public policymakers in recent years have developed behavioural insights as a tool to change water use behaviour. Behavioural interventions can be done in schools because they can have a ripple effect on the community. In addition, schools are an ideal place to do interventions because school residents are also large-scale water users [60]. A randomized control trial study in the City of Cape Town has shown that interventions on behaviour effectively reduce water usage by 15% to 26%. The interventions carried out were smart meter installation, feedback on daily and weekly water usage, and inter-school competition of water usage in 105 schools. Smart meter installation has been proven to be effective in reducing water usage at night, which indicates water usage by school staff. Meanwhile, feedback and inter-school competitions effectively decreased daytime water usage, which indicates water usage by students. Both interventions can serve as models for implementing water conservation in urban areas [52]. Then, the water-saving campaign is also being actively echoed. However, one of the study results shows that respondents still underestimate daily water use, especially to produce goods and services. As many as 85% of respondents are interested in improving local water quality. This can be a positive step to involve the community in improving water quality [61].

There are still other challenges to overcome water scarcity, that is groundwater exploitation. Groundwater exploitation can be caused by tourism, industries, and irrigation, then induce groundwater aquifer vulnerability [62]. The environmental implication due to tourism is water use. There are 10.8 thousand wells drilling in Brazil to accommodate 30–40% population's needs annually due to tourism water use. In many countries, the water consumption per person by tourists exceeds that of residents [33]. Estimating water use in Spain hotels was increased from 130 to 1,338 m^3/guest-night responsible for 16.9% of the total water footprint [63]. Moreover, the environmental impacts of the tourism sector are carbon and water footprint, aquatic eutrophication, and metals depletion. Therefore, the solution to this problem is water reuse. For instance, greywater reuse in hotels in the Spanish Mediterranean island safely reuses 23% to 80% of the total water used. It also has economic savings compared with the cost of potable water in the area. Therefore, water management is needed to regulate water use due to tourism. A study in Lloret de Mar conducted a life cycle assessment related to water reuse to address water exploitation due to tourism. The results of the study showed that water reuse can cause net carbon footprint, metals depletion, and water footprint impacts, but there are net benefits related to marine eutrophication [33].

Groundwater exploitation not only causes water scarcity but also polluted water that poses a risk to public health. A study in Dar es Salaam, Tanzania showed that industrial activities in coastal areas can put pressure on coastal aquifers that can affect groundwater quality through saltwater intrusion and chemical pollution, which become a risk of airborne diseases. As much as 88% of diarrheal diseases are caused by unsafe air supplies and poor sanitation [64]. Hence, some countries such as Europe enforce licensing related to groundwater exploitation to regulate the use of resources. However, such policies must involve community and private-sector participation [65]. To overcome the problem of water quality and quantity, various water management is needed. However, in the implementation of adequate management, there are still challenges that must be faced. The challenges are the disciplined use of groundwater; protection of aquifers and water catchment areas; and the creation of integrated and efficient technical solutions in the use of groundwater and surface water. In addition to solutions related to technical matters, programs related to water conservation are also needed that emphasize community participation, increase inspections of programs, and ensure that water management policies are truly implemented [65].

4 Understanding Health Implications of Water Contamination

Water pollution was responsible for approximately 1.36 million premature deaths in the global population in 2019. A total of 1.23 million deaths were caused by unsafe source water, and 0.76 million deaths were caused by unsafe sanitation. Women and children are the vulnerable populations who have a higher risk of dying from water pollution exposure than men [66]. Unsafe water supply affects human well-being; the infectious toxic contaminants in polluted water will burden people with diseases. The effects of acute intoxication from polluted water sometimes take years to manifest, affecting life span and long-term health impacts [67].

Waterborne or water-related diseases are those diseases that are transmitted through direct and indirect exposure, whether by ingestion or skin exposure by bathing or swimming in water contaminated with pathogens and toxic substances. Pathogens are various living organisms, usually from animal fecal, including viruses, bacteria, fungi, and intestinal worms whose presence in water is often unnoticed. Fecal–oral was the most frequent route of transmission, through drinking contaminated water or eating contaminated food that arises from poor sewage management and improper sanitation [68, 69]. The burden of diseases influenced by water pollution involves both non-communicable and communicable diseases [67]. The following examples of waterborne diseases are:

a. Diarrhea

The leading cause of diarrhea is drinking water contaminated with *Escherichia coli (E.coli)* bacteria from open defecation, dysfunctional latrines, or overflows of septic tanks or sewers [67, 69]. A study in Bangladesh revealed that household drinking water is highly contaminated with *E.coli* due to contamination at the source, inadequate treatment, unsafe storage, and unsafe water handling, which are associated with a 2.42 higher odds of diarrhea among children [70]. In 2021, World Water Development reported that 829,000 people die each year from diarrhea caused by unsafe drinking water, sanitation, and hand hygiene behavior. The severity of diarrhea in children under five years can be fatal, with more than one-third of diarrhea-related deaths occurring in this population. Diarrhea in infants and young children leads to malnutrition and a reduced immune system that increases the possibility of prolonged and recurrent diarrhea, which has a high prevalence or mortality [5].

b. Cholera

The increasing incidences of cholera and other diseases, such as trachoma, schistosomiasis, and helminthiasis, are due to the lack of appropriate water and sanitation services [5]. Cholera is caused by vibrio cholerae contamination in daily water sources [71]. Cholera can be transmitted by swimming in dirty water; however, drinking contaminated water is the leading cause of the disease transmission. People usually experience stomach ulcers, severe dehydration because of intestinal infection, rapid diarrhea and vomiting, body weakness, and sometimes cholera ends with death. Historically, cholera and typhoid are the most prominent waterborne diseases with a high prevalence of mortality [71].

c. Typhoid

Typhoid is caused by *Salmonella typhimurium* bacteria that spread by eating food or drinking water contaminated with the feces of an infected person. The *S. typhi* can live in surface water, such as sewage, drinking water, and groundwater for days [72]. Infected people commonly experience constipation or diarrhea, including high fever, with the other symptoms of typhoid are headache, stomach pain, loss of appetite, weakness, weight loss, and sometimes internal bleeding through vomiting [69, 71].

Typhoid fever can be controlled through effective water filtration. Therefore, typhoid is called a good indicator of water quality. Evidence in Pittsburgh showed that two years of water filtration can reduce roughly 75% of the death rates of typhoid [73].

d. Hepatitis A

Hepatitis A is inflammation of the liver tissue caused by viral hepatitis. The transmission is usually through the ingested contaminated water and food, commonly occurring in areas with insufficient sanitation and limited access to clean water. The Hepatitis A viruses (HAV) contaminate water by discharging untreated or improperly treated sewage. The prevalence of HAV was highest in untreated wastewater (31.4%), with the viral concentration exceeding 3.7×1010 gc/L, followed by the prevalence in groundwater (2.3%) and drinking water (0.3%). The outbreak of hepatitis A is associated with contaminated HAV in groundwater, resulting from improper well construction, nearby septic systems or sewage, heavy rainfall, and poor surface water infiltration [74]. The infections of virus hepatitis lead to jaundice in some parts of the body, loss of appetite, fatigue, and high fever [71].

e. Skin Disease

Skin diseases, mainly melanosis and keratosis, are caused by excessive arsenic contamination in water that people consume. Besides, swimming in polluted water can risk skin diseases due to exposure to various pathogens and microorganisms among swimmers [5]. Almost half of population who lived near polluted rivers in Bangladesh suffer from scabies. The people used the river for bathing, agricultural, and household purposes. Water contaminated with chemicals from industries even causes skin cancer to the human [5, 75].

f. Water-insect vector diseases and parasitic

These diseases, including malaria, dengue fever, chikungunya, zika, river blindness, yellow fever, filariasis, and other arboreal infections, are generally infectious and caused by insect vectors of intermediate hosts that breed in water [67]. Waterlogged drainage as a result of poor sanitation management constituting breeding site for the insect vectors, most of common are mosquitoes and flies. Parasites present in polluted water such as tapeworm intestinal infestation, pinworms, and roundworms (Ascaris lumbricoid), can cause abdominal pains and discomforts, retarded growth, and body weakness. There is also dracunculiasis, a crippling parasitic disease caused by *Dracunculusmedinensis*. Even rarely fatal, the diseases can make people infected non-functional for weeks.

Heavy metals released by industries harm animals, plants, and humans directly and indirectly [76]. The toxic effect of non-biodegradable heavy metals can even be impacted at the far points of the source of pollution [77]. Heavy metal pollutant such as copper, manganese, lead, cadmium, phosphate, nitrate can modify and damage several human organs like brain, kidney, lungs, liver, blood, as well caused an autoimmune disease [78]. Several heavy metals such as zinc, copper, molybdenum, nickel, manganese, and selenium are essential for plants' survival. However, they can be dangerous to human health when they cross the specific limit [76].

Pesticides are used to control plant bugs because of their lethal characteristics. Rain runoff leaches the residues into the water nearby, for instance, the nearest pond, and they become contaminated [77]. Drinking water that is contaminated with pesticides has an adverse health impact [5]. Pesticides can cause neurological problems and nervous system damage because of the carbonates and organophosphates contained [69]. Fungicides, herbicides, and insecticides are a few of the thousand forms of pesticides that contribute to water pollution. Besides contaminated water with carcinogens and other hazardous chemicals, pesticides affect ecosystems by killing weeds and negatively impacting the food chain.

Since water is essential needs of the people all over the world, the contaminated water causes serious health damage. The long-term effects of water pollution on human health often resulting in chronic diseases, development disorder, significant impact of body system, and various type of cancer. Exposure to heavy metals over a long period can dynamically prompt solid, physical, and neurological degenerative cycles of the human body, including Parkinson's disease, different sclerosis, strong dystrophy, and Alzheimer's disease [71]. Exposure to manufactured chemicals can have adverse effects on fertility and pregnancy. Reproductive disorders such as interrupted sexual development, inability to get pregnant, decreased fertility, and degraded immune function occur in those with long-term exposure to chlorides [66, 69]. Chemicals derived from plastic, including a range of toxic substances, often disrupt endocrine disorders. The symptoms of endocrine damage include diabetes, excessive hunger, excessive thirst, weight loss or gain, frequent need to urinate, swollen feet or hands, and joint aches. Apart cancer, the complication linked to long-term arsenic contamination are liver disease, diabetes, nervous system complications, such as loss of sensation in the limbs, hearing problems, and digestive difficulties [67].

Arsenic is a fatal element named the number one carcinogenic element by WHO [77]. Contamination of arsenic through consumed orally and exposure to the skin can cause skin cancer [5, 69, 79]. Kidney and bladder cancer were also the result of long-term manifestations of arsenic exposure along with trihalomethane (THM) toxic chemicals. Drinking water contaminated with nitrates levels up to 3.87 mg/L is associated with a higher risk of colorectal cancer. Lung cancer and liver damage are caused by a high concentration of Volatile Organic Compounds (VOC) in water resulting from industrialization, pesticides used, and household residues. Long-term exposure to chlorines, carcinogenic substances usually present in the water treatment process, increases the risk of liver damage and urinary and gastrointestinal cancer. The risk of several reproductive cancers, such as endometriosis, breast cancer, cervical cancer, uterine cancer, and testicular cancer, increases due to pesticide exposure in prenatal and early postnatal [66].

5 Implementing Advanced Technologies and Its Challenges

Wastewater treatment is crucial for mitigating pollution, as it removes harmful contaminants, thereby protecting human health and preserving ecosystem stability [80]. Proper treatment of wastewater involves a series of physical, chemical, and biological processes that break down pollutants into harmless substances, ensuring clean water returns to the environment [80, 81]. This not only safeguards water quality but also protects public health by removing pathogens and harmful substances that can cause infections.

Addressing these challenges necessitates the adoption of advanced treatment technologies and sustainable practices [82]. Technologies such as membrane filtration, bioreactors, and advanced oxidation processes provide innovative solutions that enhance the capability of wastewater treatment plants to handle large volumes of wastewater and remove a broad spectrum of contaminants [83]. Membrane filtration can efficiently separate suspended particles and microorganisms, while bioreactors leverage biological processes to degrade organic matter. Advanced oxidation processes utilize powerful oxidizing agents to break down persistent pollutants. Beyond technological advancements, implementing sustainable practices such as recycling treated wastewater, recovering energy from waste, and minimizing the use of harmful chemicals can further reduce the environmental impact of wastewater treatment [80, 84]. Promoting these technologies and practices requires a collaborative effort among government entities, industries, and local communities to create a comprehensive and effective approach to wastewater management [85, 86].

Sustainable practices are essential for ensuring the long-term viability of wastewater treatment systems, as they help to maintain operational efficiency and reduce environmental impacts [87]. By integrating sustainability into every aspect of wastewater management, facilities can conserve resources, lower energy consumption, and minimize waste generation. For example, adopting energy-efficient technologies and optimizing treatment processes can significantly reduce the carbon footprint of wastewater treatment plants [88]. Additionally, sustainable practices such as green infrastructure and nature-based solutions can enhance the resilience of treatment systems to climate change and other environmental stresses, ensuring that they continue to provide effective treatment even under challenging conditions [89]. Ultimately, these practices contribute to the sustainability of water resources, protecting both the environment and public health for future generations [87].

Sustainable practices in wastewater management include a range of strategies aimed at reducing environmental impacts and conserving resources. One such practice is reducing water consumption, which involves the use of water-saving technologies and public education to promote efficient water use, thereby decreasing the volume of wastewater that needs to be treated [87, 90]. Another crucial practice is recycling treated wastewater, which entails treating wastewater to a high enough standard that it can be safely reused for various applications, such as agricultural irrigation, industrial processes, or even as a supplement to potable water supplies. This practice not only reduces the strain on freshwater resources but also helps to create a more circular water economy. Additionally, minimizing the use of harmful chemicals in treatment processes is vital for protecting both the environment and human health [77]. This can be achieved by adopting alternative treatment methods, such as biological processes, natural filtration systems, and the use of environmentally friendly disinfectants, which reduce the release of toxic substances into the environment.

Implementing advanced wastewater treatment technologies presents several challenges, primarily due to the high initial capital costs associated with acquiring and installing sophisticated equipment. For instance, technologies like membrane filtration and advanced oxidation processes require substantial financial investment upfront, which can be a significant barrier for many municipalities and industries [91]. Additionally, the technical complexity of these systems necessitates specialized knowledge and expertise for their operation and maintenance. This includes the need for trained personnel who can manage the intricacies of advanced processes, monitor system performance, and troubleshoot issues as they arise. Without sufficient technical expertise, the risk of operational failures increases, potentially leading to higher maintenance costs and reduced efficiency [92]. These challenges highlight the importance of addressing financial and technical barriers to facilitate the widespread adoption of advanced wastewater treatment technologies. These challenges can be addressed through government support, such as subsidies and grants to offset initial costs, industry collaboration to share expertise and streamline technology adoption, and public awareness campaigns to educate communities on the benefits and importance of advanced wastewater treatment technologies [87]. Investments in research and development are essential to enhance the feasibility and affordability of advanced wastewater treatment technologies. R&D can lead to innovations that reduce costs and improve efficiency, making these technologies more accessible and practical for widespread use [93].

6 Collaborative Wastewater Management

Wastewater management is removing contaminants from wastewater so that effluent can be returned to the water cycle. Wastewater management faces several challenges. The challenges discussed, such as water scarcity, pollution control, and sustainable resource management, highlight the urgency of adopting proactive measures. By harnessing emerging technologies and optimizing processes, industries can mitigate the adverse effects of industrial wastewater discharge on ecosystems and public health [31, 86, 87]. While challenges are significant, the enablers identified present promising opportunities for progress. Policy support, stakeholder engagement, technological advancements, and sustainable practices collectively pave the way for a more responsible industrial wastewater management paradigm. It is crucial for industries to prioritize environmentally conscious practices, not just to comply with regulations but also to enhance societal well-being and environmental health [94].

The inclusion of water issues in the Sustainable Development Goals (SDGs) in 2015 underscores global efforts towards clean water and sanitation (SDG6). Stakeholders across policymakers, NGOs, businesses, academia, and society have contributed to integrating sustainable development principles. In addressing these challenges, it is essential to adopt corresponding water management policies to reduce the harm caused by water pollution to human health. Additionally, health education should be widely carried out to raise awareness and promote best practices in water management. Health education is conducted through various channels, such as introducing environmental education, educating residents about clean water, monitoring clean water, and managing domestic wastewater through newspapers, magazines, television, and the Internet to enhance public health awareness. By promoting awareness and encouraging responsible practices, communities can actively contribute to improving water quality [5]. Educational programs play a crucial role in Germany's successful waste management by raising awareness and promoting sustainable practices. The municipal company AWM in Münster collected 130,557 tons of domestic waste in 2018, with a comprehensive collection system. These initiatives combined with PAYT (Pay-As-You-Throw) fees, have significantly contributed to high recycling rates, exemplified by Münster's 67% recycling rate for household waste [95].

Besides the importance of education and raising awareness regarding water pollution and wastewater management, there are substantial factors involved in the implementation of government regulatory policies, as seen in Europe. Government policies and regulations in Europe, exemplified by directives like the Waste Framework Directive and Water Framework Directive, play a critical role in shaping effective wastewater and waste management practices. The EU has made significant strides in waste management, with increasing recycling rates for municipal waste rising from 37% in 2008 to 46% in 2016 under the Circular Economy Action Plan. Despite progress, disparities among Member States and waste streams persist. Water quality improvements under directives like the Urban Waste Water Treatment and Drinking Water Directives are notable with 85% of inspected sites meeting stringent standards by 2017, challenges remain in achieving overall water body health. The shift towards a circular economy model necessitates sustainable practices in water management, emphasizing pollutant reclamation and resource recovery, thereby advancing Europe's sustainability goals. This holistic approach aims to transform Europe's economy towards sustainability further [96–98]. These studies indicate that government policies and regulations in a country significantly influence the implementation of wastewater management within the framework. Therefore, a framework for wastewater management should be established as the initial plan for wastewater management.

Furthermore, the role of government bodies in coordinating stakeholders is crucial, especially through Public–public partnerships (PuPs) in wastewater treatment, where various public sector entities collaborate to ensure effective provision and management of wastewater services. These arrangements are designed to leverage the strengths and resources of multiple public bodies to achieve greater efficiency, cost savings, and service quality compared to other models like private sector participation. Key aspects of PuPs include horizontal integration (cooperation between public entities at the same governmental level, such as municipalities working together), vertical integration (collaboration between different levels of government, for example, local and national government agencies, scale economies (cost savings achieved by pooling resources and increasing the scale of operations), operational efficiency (improved efficiency due to shared expertise, resources, and streamlined processes), trust and reciprocity (strong cooperation based on mutual trust and shared goals, reducing opportunistic behaviors). In wastewater management, these partnerships can help in financing, constructing, and operating facilities more effectively than individual public entities or through private sector partnerships. Studies on public–public partnerships (PPPs) reveal varied impacts such as Brazil and Sweden benefit from scale economies and trust; South Africa reduces costs through cooperation; China employs BOT contracts for reduced risk; Portugal faces higher costs due to contractual challenges in wastewater management highlighting global lessons for effective public sector collaborations [99].

In addition to the role of governments in implementing regulations or policies, technology plays a crucial role in wastewater management. Technological innovations in wastewater management, including integrated approaches and advanced technologies play a pivotal role in enhancing asset management and preventing pollution. Circular bio-economy practices contribute significantly by reducing waste and resource consumption, aligning with sustainable development goals. AI and advanced tools further optimize treatment efficiency, advancing environmental management efforts. Effective urban wastewater management necessitates considerations such as predicting water demand, detecting contamination, evaluating sewer defects, and monitoring assets to mitigate urban flooding and maintain wastewater system health. Promising methods like modified electrochemical oxidation, gamma radiation, and plasma-assisted systems demonstrate sustainability and efficiency in contaminant conversion. Biotechnological solutions, such as algal wastewater treatment plants, offer cost-effective treatment with a reduced carbon footprint. Additionally, methods like groundwater tracers, MXene-based nanomaterials, as nanomaterials can be customized to efficiently purify water by leveraging their tailored properties [100], bismuth-based photo-catalysts and effect-based techniques provide effective alternatives for comprehensive wastewater management strategies [94].

One approach to wastewater management can be carried out by wastewater reuse. The global trend towards wastewater reuse is increasingly evident as countries worldwide adopt regulations to address water scarcity and environmental sustainability. Alcalde-Sanz and Gawlik highlight that countries like Australia, Canada, China, Israel, Japan, Jordan, Mexico, South Africa, Tunisia, the USA, and various EU states have implemented specific criteria for wastewater reuse. In Europe, pressures from climate change and urbanization have strained freshwater resources, prompting efforts to enhance water reuse. Despite this, only six EU member states have established regulatory standards for wastewater reuse, primarily focusing on agricultural irrigation. Recently, the European Commission proposed new standards categorizing reclaimed water quality based on crop type, irrigation method, and treatment levels, aligning with WHO guidelines. France stands out among European nations with wastewater reuse standards dating back to 1991, adhering closely to WHO guidelines and imposing additional restrictions. Jordan emerged as a leader in wastewater reuse due to severe water scarcity, where 90% of treated wastewater is reused for agricultural irrigation, reducing fertilizer use significantly and yielding economic benefits. Comparative analysis reveals wide variations in wastewater reuse standards globally, with differing limits and applications tailored to crop and irrigation specifics. The EU's evolving regulations indicate a comprehensive approach to safety standards, contrasting with less stringent practices observed in other regions like India. Overall, these efforts underscore the diverse strategies countries employ to manage wastewater effectively amidst global environmental challenges [101].

The supportive governmental policies are leading to investments supporting the rapid development of new technologies and approaches to integrated water/wastewater management which favor the implementation of wastewater treatment and reclamation [83, 87, 91]. Good governance should help make investment in water and sanitation services more equitable, targeted, and sustainable. This involves managing systems to ensure continuous service delivery to those in need and allowing for expansion as the population grows and new demands arise. Poor management, often resulting from an unsupportive governance framework, leads to reduced service coverage as systems deteriorate. Therefore, investing in institutional capacity may be more efficient for achieving sustained growth in service coverage compared to direct investment in physical infrastructure.

7 Conclusion and Recommendation

Access to clean water, adequate sanitation, and proper hygiene is vital for human health, yet increasing global water pollution from industrial and agricultural activities, coupled with growing water consumption, presents significant challenges. Approximately 2 billion people lack safe drinking water, with 80% of industrial waste discharged untreated into the environment, leading to severe health risks from toxic substances like heavy metals. Groundwater depletion, exacerbated by excessive freshwater withdrawal and industrial activities, further threatens water availability and quality, heightening the risk of water scarcity and pollution. Effective solutions require adopting water-efficient technologies in industries, implementing strict government policies on wastewater treatment, fostering public awareness about responsible water use, and promoting community-based water management systems. Additionally, regulatory measures such as licensing for groundwater exploitation and community participation in water governance are crucial. Sustainable water management practices, including behavioral interventions like water-saving campaigns and smart meter installations, must be prioritized to address water pollution, ensure water security, and protect public health.

References

1. Mateo-Sagasta J, Marjani Zadeh S, Turral H (2017) Water pollution from agriculture: A global review. Executive summary. Food and Agriculture Organization of the United Nations
2. Ritchie H, Spooner F, Roser M (2023) Clean water and sanitation. https://ourworldindata.org/clean-water-sanitation
3. Alijanzadeh Maliji B, Babayeemehr A (2023) Department of Law, Faculty of Law, Chalous Branch, Islamic Azad University, Chalous, Iran., Mehrabani, S., Non-Communicable Pediatric Disease Research Center, Health Research Institute, Babol University of Medical Sciences, Babol, Iran., Aghajanpour, F., & The Clinical Research Development Unit of Amirkola Children's Hospital, Student Research Committee, Babol University of Medical Sciences, Babol, Iran. (2023). Role of the world health organization in management of gastrointestinal diseases caused by contaminated water in children in the Middle East: A review article. J Pediatr Rev 11(1):59–66. https://doi.org/10.32598/jpr.11.1.285.2
4. Chen B, Wang M, Duan M, Ma X, Hong J, Xie F, Zhang R, Li X (2019) In search of key: Protecting human health and the ecosystem from water pollution in China. J Clean Prod 228:101–111. https://doi.org/10.1016/j.jclepro.2019.04.228
5. Lin L, Yang H, Xu X (2022) Effects of water pollution on human health and disease heterogeneity: A review. Front Environ Sci 10:880246. https://doi.org/10.3389/fenvs.2022.880246
6. Wu H, Gai Z, Guo Y, Li Y, Hao Y, Lu Z-N (2020) Does environmental pollution inhibit urbanization in China? A new perspective through residents' medical and health costs. Environ Res 182:109128. https://doi.org/10.1016/j.envres.2020.109128
7. Syafrudin M, Kristanti RA, Yuniarto A, Hadibarata T, Rhee J, Al-onazi WA, Algarni TS, Almarri AH, Al-Mohaimeed AM (2021) Pesticides in drinking water—A review. Int J Environ Res Public Health 18(2):468. https://doi.org/10.3390/ijerph18020468
8. Sadoff CW, Hall JW, Grey D, Aerts JC (2015) Securing Water, Sustaining Growth: Report of the GWP/OECD Task Force on Water Security and Sustainable Growth. University of Oxford
9. Schlosser CA, Strzepek K, Gao X, Fant C, Blanc É, Paltsev S, Jacoby H, Reilly J, Gueneau A (2014) The future of global water stress: An integrated assessment. Earth's Future 2(8):341–361. https://doi.org/10.1002/2014EF000238
10. Jasechko S, Perrone D (2021) Global groundwater wells at risk of running dry. Science 372(6540):418–421. https://doi.org/10.1126/science.abc2755
11. Rodell M, Famiglietti JS, Wiese DN, Reager JT, Beaudoing HK, Landerer FW, Lo M-H (2018) Emerging trends in global freshwater availability. Nature 557(7707):651–659. https://doi.org/10.1038/s41586-018-0123-1
12. Scanlon BR, Fakhreddine S, Rateb A, De Graaf I, Famiglietti J, Gleeson T, Grafton RQ, Jobbagy E, Kebede S, Kolusu SR, Konikow LF, Long D, Mekonnen M, Schmied HM, Mukherjee A, MacDonald A, Reedy RC, Shamsudduha M, Simmons CT, Zheng C (2023) Global water resources and the role of groundwater in a resilient water future. Nat Rev Earth & Environ 4(2):87–101. https://doi.org/10.1038/s43017-022-00378-6
13. Alfarrah N, Walraevens K (2018) Groundwater overexploitation and seawater intrusion in coastal areas of arid and semi-arid regions. Water 10(2):143. https://doi.org/10.3390/w10020143
14. Du J, Laghari Y, Wei Y-C, Wu L, He A-L, Liu G-Y, Yang H-H, Guo Z-Y, Leghari SJ (2024) Groundwater depletion and degradation in the north china plain: Challenges and mitigation options. Water 16(2):354. https://doi.org/10.3390/w16020354
15. Mishra RK (2023) Fresh water availability and its global challenge. Br J Multidiscip Adv Stud 4(3):1–78. https://doi.org/10.37745/bjmas.2022.0208
16. Lu Y, Song S, Wang R, Liu Z, Meng J, Sweetman AJ, Jenkins A, Ferrier RC, Li H, Luo W, Wang T (2015) Impacts of soil and water pollution on food safety and health risks in China. Environ Int 77:5–15. https://doi.org/10.1016/j.envint.2014.12.010

17. Calaf GM (2021) Role of organophosphorous pesticides and acetylcholine in breast carcinogenesis. Semin Cancer Biol 76:206–217. https://doi.org/10.1016/j.semcancer.2021.03.016
18. Kalyabina VP, Esimbekova EN, Kopylova KV, Kratasyuk VA (2021) Pesticides: Formulants, distribution pathways and effects on human health—a review. Toxicol Rep 8:1179–1192. https://doi.org/10.1016/j.toxrep.2021.06.004
19. Pathak VM, Verma VK, Rawat BS, Kaur B, Babu N, Sharma A, Dewali S, Yadav M, Kumari R, Singh S, Mohapatra A, Pandey V, Rana N, Cunill JM (2022) Current status of pesticide effects on environment, human health and it's eco-friendly management as bioremediation: A comprehensive review. Front Microbiol 13:962619. https://doi.org/10.3389/fmicb.2022.962619
20. Shrestha A, Kunwar BM, Meierhofer R (2022) Water, sanitation, hygiene practices, health and nutritional status among children before and during the COVID-19 pandemic: Longitudinal evidence from remote areas of Dailekh and Achham districts in Nepal. BMC Public Health 22(1):2035. https://doi.org/10.1186/s12889-022-14346-8
21. Chen Y, Li C, Li X, Zhang X, Tan Q (2022) Efficiency of water pollution control based on a three-stage SBM-DEA model. Water 14(9):1453. https://doi.org/10.3390/w14091453
22. Nti EK, Cobbina SJ, Attafuah EE, Senanu LD, Amenyeku G, Gyan MA, Forson D, Safo A-R (2023) Water pollution control and revitalization using advanced technologies: Uncovering artificial intelligence options towards environmental health protection, sustainability and water security. Heliyon 9(7):e18170. https://doi.org/10.1016/j.heliyon.2023.e18170
23. Khatibi FS, Dedekorkut-Howes A, Howes M, Torabi E (2021) Can public awareness, knowledge and engagement improve climate change adaptation policies? Discover Sustainability 2(1):18. https://doi.org/10.1007/s43621-021-00024-z
24. Sheehan WJ, Rangsithienchai PA, Wood RA, Rivard D, Chinratanapisit S, Perzanowski MS, Chew GL, Seltzer JM, Matsui EC, Phipatanakul W (2010) Pest and allergen exposure and abatement in inner-city asthma: A work group report of the American academy of allergy, asthma & immunology indoor allergy/air pollution committee. J Allergy Clin Immunol 125(3):575–581. https://doi.org/10.1016/j.jaci.2010.01.023
25. Howell CL, Cortado AP, Ünver O (2023) Stakeholder engagement and perceptions on water governance and water management in Azerbaijan. Water 15(12):2201. https://doi.org/10.3390/w15122201
26. Okumah M, Yeboah AS, Amponsah O (2020) Stakeholders' willingness and motivations to support sustainable water resources management: Insights from a Ghanaian study. Conserv Sci Pract 2(3):e170. https://doi.org/10.1111/csp2.170
27. Wang J, Zhou W, Zhao M, Guo X (2023) Water quality assessment and pollution evaluation of surface water sources: The case of Weishan and Luoma Lakes, Xuzhou, Jiangsu Province China. Environ Technol Innov 32:103397. https://doi.org/10.1016/j.eti.2023.103397
28. Basu NB, Van Meter KJ, Byrnes DK, Van Cappellen P, Brouwer R, Jacobsen BH, Jarsjö J, Rudolph DL, Cunha MC, Nelson N, Bhattacharya R, Destouni G, Olsen SB (2022) Managing nitrogen legacies to accelerate water quality improvement. Nat Geosci 15(2):97–105. https://doi.org/10.1038/s41561-021-00889-9
29. Mao G, Han Y, Liu X, Crittenden J, Huang N, Ahmad UM (2022) Technology status and trends ofindustrial wastewater treatment: a patent analysis. Chemosphere 288. https://doi.org/10.1016/j.chemosphere.2021.132483
30. WWAP (2018) The united nations world water development report 2018: nature-based solutions for water. InUN water report. https://digitallibrary.un.org/record/3893097?v=pdf
31. Zacchaeus OO, Adeyemi MB, Azeem A, Adegoke KA, Anumah AO, Taiwo AM, Ganiyu SA (2020) Effects of industrialization on groundwater quality in Shagamu and Ota industrial areas of Ogun state Nigeria. Heliyon 6(7):e04353. https://doi.org/10.1016/j.heliyon.2020.e04353
32. Kumar V, Parihar RD, Sharma A, Bakshi P, Singh Sidhu GP, Bali AS, Karaouzas I, Bhardwaj R, Thukral AK, Gyasi-Agyei Y, Rodrigo-Comino J (2019) Global evaluation of heavy metal content in surface water bodies: A meta-analysis using heavy metal pollution indices and multivariate statistical analyses. Chemosphere 236:124364. https://doi.org/10.1016/j.chemosphere.2019.124364

33. Santana CS, Montalván Olivares DM, Silva VHC, Luzardo FHM, Velasco FG, De Jesus RM (2020) Assessment of water resources pollution associated with mining activity in a semi-arid region. J Environ Manage 273:111148. https://doi.org/10.1016/j.jenvman.2020.111148
34. Li S, Yu L, Jiang W, Yu H, Wang X (2022) The recent progress china has made in green mine construction, part I: Mining groundwater pollution and sustainable mining. Int J Environ Res Public Health 19(9):5673. https://doi.org/10.3390/ijerph19095673
35. Petronijević N, Radovanović D, Štulović M, Sokić M, Jovanović G, Kamberović Ž, Stanković S, Stopic S, Onjia A (2022) Analysis of the mechanism of acid mine drainage neutralization using fly ash as an alternative material: A case study of the extremely acidic lake Robule in Eastern Serbia. Water 14(20):3244. https://doi.org/10.3390/w14203244
36. Vidal P, Martínez-Martínez M, Fernandez-Lopez L, Roda S, Méndez-García C, Golyshina OV, Guallar V, Peláez AI, Ferrer M (2022) Metagenomic mining for esterases in the microbial community of Los Rueldos acid mine drainage formation. Front Microbiol 13:868839. https://doi.org/10.3389/fmicb.2022.868839
37. Turingan COA, Cordero KS, Santos AL, Tan GSL, Tabelin CB, Alorro RD, Orbecido AH (2022) Acid mine drainage treatment using a process train with laterite mine waste, concrete waste, and limestone as treatment media. Water 14(7):1070. https://doi.org/10.3390/w14071070
38. Rakotondrabe F, Ndam JR, Mfonka Z, Rasolomanana EH, Nyangono AJ, Ako A (2018) Water quality assessment in the Bétaré-Oya gold mining area (East-Cameroon): Multivariate statistical analysis approach. Sci Total Environ 610–611:831–844. https://doi.org/10.1016/j.scitotenv.2017.08.080
39. Mateo-Sagasta J, Marjani S, Turral H (2018) More people, more food, worse water?: A global review of water pollution from agriculture. Food and Agriculture Organization of the United Nations (FAO) and International Water Management Institute
40. Sahoo PK, Kim K, Powell MA (2016) Managing groundwater nitrate contamination from livestock farms: implication for nitrate management guidelines. Curr Pollut Rep 2(3):178–187. https://doi.org/10.1007/s40726-016-0033-5
41. Aranguren-Díaz Y, Galán-Freyle NJ, Guerra A, Manares-Romero A, Pacheco-Londoño LC, Romero-Coronado A, Vidal-Figueroa N, Machado-Sierra E (2024) Aquifers and groundwater: Challenges and opportunities in water resource management in Colombia. Water 16(5):685. https://doi.org/10.3390/w16050685
42. Abedzadeh S, Roozbahani A, Heidari A (2020) Risk assessment of water resources development plans using fuzzy fault tree analysis. Water Resour Manage 34(8):2549–2569. https://doi.org/10.1007/s11269-020-02578-5
43. Ducci L, Rizzo P, Pinardi R, Solfrini A, Maggiali A, Pizzati M, Balsamo F, Celico F (2022) What is the impact of leaky sewers on groundwater contamination in urban semi-confined aquifers? A test study related to Fecal matter and personal care products (PCPs). Hydrology 10(1):3. https://doi.org/10.3390/hydrology10010003
44. Teixeira CA, Guerra M (2024) Municipal solid waste—addressing environmental concerns. Sustainability 16(3):1235. https://doi.org/10.3390/su16031235
45. Bossa A, Diekkrüger B, Agbossou E (2014) Scenario-based impacts of land use and climate change on land and water degradation from the Meso to regional scale. Water 6(10):3152–3181. https://doi.org/10.3390/w6103152
46. Jha PK, Tripathi P (2021) Arsenic and fluoride contamination in groundwater: a review of global scenarioswith special reference to India. Groundw Sustain Dev 13(100576). https://doi.org/10.1016/j.gsd.2021.100576
47. Sahu P (2019) Fluoride pollution in groundwater. In: Sikdar PK (Ed) Groundwater Development and Management (pp 329–350). Springer International Publishing. https://doi.org/10.1007/978-3-319-75115-3_14
48. Hecht H, Oguchi T (2017) Global evaluation of erosion rates in relation to tectonics. Prog Earth Planet Sci 4(1):40. https://doi.org/10.1186/s40645-017-0156-3
49. Sharpley A, Jarvie HP, Buda A, May L, Spears B, Kleinman P (2013) Phosphorus legacy: Overcoming the effects of past management practices to mitigate future water quality impairment. J Environ Qual 42(5):1308–1326. https://doi.org/10.2134/jeq2013.03.0098

50. Bradshaw JK, Snyder BJ, Oladeinde A, Spidle D, Berrang ME, Meinersmann RJ, Oakley B, Sidle RC, Sullivan K, Molina M (2016) Characterizing relationships among fecal indicator bacteria, microbial source tracking markers, and associated waterborne pathogen occurrence in stream water and sediments in a mixed land use watershed. Water Res 101:498–509. https://doi.org/10.1016/j.watres.2016.05.014
51. WHO (2019) Microplastics in drinking-water. WHO
52. Visser M, Booysen M, Berger K, Brühl J (2019) Saving water at Cape Town schools by using smart metering and behavioural change. https://doi.org/10.31224/osf.io/7kzwv
53. Jiménez A, Saikia P, Giné R, Avello P, Leten J, Liss B, Schneider K, Ward R (2020) Unpacking water governance: A framework for practitioners. Water 12(3):827. https://doi.org/10.3390/w12030827
54. Hove J, D'Ambruoso L, Mabetha D, Van Der Merwe M, Byass P, Kahn K, Khosa S, Witter S, Twine R (2019) 'Water is life': Developing community participation for clean water in rural South Africa. BMJ Glob Health 4(3):e001377. https://doi.org/10.1136/bmjgh-2018-001377
55. Behnke NL, Klug T, Cronk R, Shields KF, Lee K, Kelly ER, Allgood G, Bartram J (2017) Resource mobilization for community-managed rural water systems: Evidence from Ghana, Kenya, and Zambia. J Clean Prod 156:437–444. https://doi.org/10.1016/j.jclepro.2017.04.016
56. Hutchings P, Franceys R, Mekala S, Smits S, James AJ (2017) Revisiting the history, concepts and typologies of community management for rural drinking water supply in India. Int J Water Resour Dev 33(1):152–169. https://doi.org/10.1080/07900627.2016.1145576
57. Truslove JPVM, Miller A, Mannix N, Nhlema M, Rivett MO, Coulson AB, Mleta P, Kalin RM (2019) Understanding the functionality and burden on decentralised rural water supply: influence of millennium development goal 7c coverage targets. Water 11(3):494. https://doi.org/10.3390/w11030494
58. Machado AVM, Oliveira PAD, Matos PG (2022) Review of community-managed water supply—factors affecting its long-term sustainability. Water 14(14):2209. https://doi.org/10.3390/w14142209
59. Adams EA, Stoler J, Adams Y (2020) Water insecurity and urban poverty in the Global South: Implications for health and human biology. Am J Hum Biol 32(1):e23368. https://doi.org/10.1002/ajhb.23368
60. Booysen M, Ripunda C, Visser M (2019) Results from a water-saving maintenance campaign at Cape Town schools in the run-up to Day Zero. https://doi.org/10.31224/osf.io/45cf9
61. Seelen LMS, Flaim G, Jennings E, De Senerpont LN (2019) Saving water for the future: Public awareness of water usage and water quality. J Environ Manage 242:246–257. https://doi.org/10.1016/j.jenvman.2019.04.047
62. Li H, Yu X, Zhang W, Huan Y, Yu J, Zhang Y (2018) Risk assessment of groundwater organic pollution using hazard, intrinsic vulnerability, and groundwater value, Suzhou city in China. Exposure and Health 10(2):99–115. https://doi.org/10.1007/s12403-017-0248-8
63. Becken S, McLennan C (2017) Evidence of the water-energy nexus in tourist accommodation. J Clean Prod 144:415–425. https://doi.org/10.1016/j.jclepro.2016.12.167
64. Mussa K, Mjemah I, Walraevens K (2019) Quantification of groundwater exploitation and assessment of water quality risk perception in the Dar Es Salaam quaternary Aquifer Tanzania. *Water* 11(12):2552. https://doi.org/10.3390/w11122552
65. Conicelli B, Hirata R, Galvão P, Aranda N, Terada R, Gutiérrez OJG (2021) Groundwater governance: The illegality of exploitation and ways to minimize the problem. An Acad Bras Ciênc 93(1):e20200623. https://doi.org/10.1590/0001-3765202120200623
66. Fuller R, Landrigan PJ, Balakrishnan K, Bathan G, Bose-O'Reilly S, Brauer M, Caravanos J, Chiles T, Cohen A, Corra L, Cropper M, Ferraro G, Hanna J, Hanrahan D, Hu H, Hunter D, Janata G, Kupka R, Lanphear B, Yan C (2022) Pollution and health: A progress update. The Lancet Planetary Health 6(6):e535–e547. https://doi.org/10.1016/S2542-5196(22)00090-0
67. Boelee E, Geerling G, Van Der Zaan B, Blauw A, Vethaak AD (2019) Water and health: From environmental pressures to integrated responses. Acta Trop 193:217–226. https://doi.org/10.1016/j.actatropica.2019.03.011

68. Astuti RDP, Mallongi A, Choi K, Amiruddin R, Hatta M, Tantrakarnapa K, Rauf AU (2022) Health risks from multiroute exposure of potentially toxic elements in a coastal community: A probabilistic risk approach in Pangkep Regency, Indonesia. Geomat Nat Haz Risk 13(1):705–735. https://doi.org/10.1080/19475705.2022.2041110
69. Rehman B, Hassan TU, Bano A (2019) Potential of indole-3-acetic acid-producing rhizobacteria to resist Pb toxicity in polluted soil. Soil Sediment Contam: Int J 28(1):101–121. https://doi.org/10.1080/15320383.2018.1539947
70. Khan JR, Hossain MdB, Chakraborty PA, Mistry SK (2022) Household drinking water E. coli contamination and its associated risk with childhood diarrhea in Bangladesh. Environ Sci Pollut Res 29(21):32180–32189. https://doi.org/10.1007/s11356-021-18460-9
71. Stanley IC, Njoku VON, Arinze C, Chizoruo IF, Blessing EN (2021) A Review: Effects of air, water and land dumpsite on human health and analytical methods for determination of pollutants. Anal Methods Environ Chem J 4(03):80–106. https://doi.org/10.24200/amecj.v4.i03.147
72. Kaluse PS, Bhatt N, Bankar N (2021) Study of typhoid fever: A review. J Pharm Res Int 286–291. https://doi.org/10.9734/jpri/2021/v33i39A32172
73. Beach B, Ferrie J, Saavedra M, Troesken W (2016) Typhoid fever, water quality, and human capital formation. J Econ Hist 76(1):41–75. https://doi.org/10.1017/S0022050716000413
74. Takuissu GR, Kenmoe S, Ebogo-Belobo JT, Kengne-Ndé C, Mbaga DS, Bowo-Ngandji A, Ndzie Ondigui JL, Kenfack-Momo R, Tchatchouang S, Kenfack-Zanguim J, Lontuo Fogang R, Zeuko'o Menkem E, Kame-Ngasse GI, Magoudjou-Pekam JN, Veneri C, Mancini P, Bonanno Ferraro G, Iaconelli M, Orlandi L, La Rosa G (2023) Occurrence of hepatitis a virus in water matrices: a systematic review and meta-analysis. Int J Environ Res Public Health 20(2):1054. https://doi.org/10.3390/ijerph20021054
75. Hanif M, Miah R, Islam M, Marzia S (2020) Impact of Kapotaksha river water pollution on human health and environment. Progress Agric 31(1):1–9. https://doi.org/10.3329/pa.v31i1.48300
76. Arif A (2020) Water pollution and industries. Pure Appl Biol 9(4). https://doi.org/10.19045/bspab.2020.90237
77. Hasan MdK, Shahriar A, Jim KU (2019) Water pollution in Bangladesh and its impact on public health. Heliyon 5(8):e02145. https://doi.org/10.1016/j.heliyon.2019.e02145
78. Chukwuemeka I, Njoku V, Arinze C, Chizoruo I, Blessing E (2021) A review: effects of air, water and landdumpsite on human health and analytical methods for determination of pollutants. Anal Methods Environ Chem J4(3):480–106. https://doi.org/10.24200/amecj.v4.i03.147
79. Babuji P, Thirumalaisamy S, Duraisamy K, Periyasamy G (2023) Human health risks due to exposure to water pollution: A review. Water 15(14):2532. https://doi.org/10.3390/w15142532
80. Ugrina M, Milojković J (2024) Advances in wastewater treatment. Energies 17(6):1400. https://doi.org/10.3390/en17061400
81. Aghalari Z, Dahms H-U, Sillanpää M, Sosa-Hernandez JE, Parra-Saldívar R (2020) Effectiveness of wastewater treatment systems in removing microbial agents: A systematic review. Glob Health 16(1):13. https://doi.org/10.1186/s12992-020-0546-y
82. Garg A, Gautamb P, Salunkeb D (2022) Advanced treatment technologies for industrial wastewater. In: Roy S, Garg A, Garg S, Tran TA (Eds) Advanced Industrial Wastewater Treatment and Reclamation of Water (pp 25–44). Springer International Publishing. https://doi.org/10.1007/978-3-030-83811-9_2
83. Obotey E, Rathilal S (2020) Membrane technologies in wastewater treatment: A review. Membranes 10(5):89. https://doi.org/10.3390/membranes10050089
84. Bhatti S, Richards R, McGinn P (2021) Screening of two freshwater green microalgae in pulp and paper mill wastewater effluents in Nova Scotia Canada. Water Sci Technol 83(6):1483–1498. https://doi.org/10.2166/wst.2021.001
85. Rayaroth MP, Boczkaj G, Aubry O, Aravind UK, Aravindakumar CT (2023) Advanced oxidation processes for degradation of water pollutants—Ambivalent impact of carbonate species: A review. Water 15(8):1615. https://doi.org/10.3390/w15081615

86. Śniatała B, Al-Hazmi HE, Sobotka D, Zhai J, Mąkinia J (2024) Advancing sustainable wastewater management: A comprehensive review of nutrient recovery products and their applications. Sci Total Environ 937:173446. https://doi.org/10.1016/j.scitotenv.2024.173446
87. Silva JA (2023) Wastewater treatment and reuse for sustainable water resources management: A systematic literature review. Sustainability 15(14):10940. https://doi.org/10.3390/su151410940
88. Moktadir MdA, Ren J, Zhou J (2023) A systematic review on tannery sludge to energy route: Current practices, impacts, strategies, and future directions. Sci Total Environ 901:166244. https://doi.org/10.1016/j.scitotenv.2023.166244
89. Fang C, Awoyemi OS, Saianand G, Xu L, Niu J, Naidu R (2024) Characterising microplastics in indoor air: Insights from Raman imaging analysis of air filter samples. J Hazard Mater 464:132969. https://doi.org/10.1016/j.jhazmat.2023.132969
90. Bouramdane A-A (2023) Optimal water management strategies: paving the way for sustainability in smart cities. Smart Cities 6(5):2849–2882. https://doi.org/10.3390/smartcities6050128
91. Siagian UWR, Khoiruddin K, Wardani AK, Aryanti PTP, Widiasa IN, Qiu G, Ting YP, Wenten IG (2021) High-performance ultrafiltration membrane: recent progress and its application for wastewater treatment. Curr Pollut Rep 7(4):448–462. https://doi.org/10.1007/s40726-021-00204-5
92. Weng C-H (2020) Water pollution prevention and state of the art treatment technologies. Environ Sci Pollut Res 27(28):34583–34585. https://doi.org/10.1007/s11356-020-09994-5
93. Constance Obiuto N, David Ugwuanyi E, Ninduwezuor-Ehiobu N, Chigozie Ani E, Andrew Olu-lawal K (2024) Advancing wastewater treatment technologies: The role of chemical engineering simulations in environmental sustainability. World J Adv Res Rev 21(3):019–031. https://doi.org/10.30574/wjarr.2024.21.3.0649
94. Singh BJ, Chakraborty A, Sehgal R (2023) A systematic review of industrial wastewater management: Evaluating challenges and enablers. J Environ Manage 348:119230. https://doi.org/10.1016/j.jenvman.2023.119230
95. Azevedo BD, Scavarda LF, Caiado RGG, Fuss M (2021) Improving urban household solid waste management in developing countries based on the German experience. Waste Manage 120:772–783. https://doi.org/10.1016/j.wasman.2020.11.001
96. Deselnicu DC, Militaru G, Deselnicu V, Zăinescu G, Albu L (2018) Towards a Circular Economy– a Zero Waste Programme for Europe. 563–568. https://doi.org/10.24264/icams-2018.XI.4
97. Rauf AU (2024) Electronic waste problem in developing nations: Mismanagement, health implications, and circular economy opportunities. J Kesehat Lingkung 16(1):18–31. https://doi.org/10.20473/jkl.v16i1.2024.18-31
98. Smol M, Adam C, Preisner M (2020) Circular economy model framework in the European water and wastewater sector. J Mater Cycles Waste Manage 22(3):682–697. https://doi.org/10.1007/s10163-019-00960-z
99. Silvestre HC, Marques RC, Gomes RC (2018) Joined-up Government of utilities: A meta-review on a public–public partnership and inter-municipal cooperation in the water and wastewater industries. Public Manag Rev 20(4):607–631. https://doi.org/10.1080/14719037.2017.1363906
100. Pérez H, Quintero García OJ, Amezcua-Allieri MA, Rodríguez Vázquez R (2023) Nanotechnology as an efficient and effective alternative for wastewater treatment: An overview. Water Sci Technol 87(12):2971–3001. https://doi.org/10.2166/wst.2023.179
101. Schellenberg T, Subramanian V, Ganeshan G, Tompkins D, Pradeep R (2020) Wastewater discharge standards in the evolving context of urban sustainability-the case of India. Front Environ Sci 8:30. https://doi.org/10.3389/fenvs.2020.00030

Air Quality and Urban Sustainable Development- Current Issues and Future Directions

Ari Prayogo Pribadi, Annisa Utami Rauf, Yusuf Mukasyafah Rizqi Rahman, and Zilfani Fuadiyah Haq

Abstract Air quality and urban sustainability are highly interrelated themes, have overarching impacts on public health, environmental quality, resource fulfillment and social justice. Understanding the inter-relationships of air quality with urban sustainable development is crucial in framing runs further. Urban areas are growing fast worldwide, and the quality of urban air is one key issue in city residents' health, well-being and economy. To design effective policies and interventions to ensure clean air, mitigate pollution, attain improved public health outcomes and derive sustainable urban growth it is critical for us to comprehend at least some of this intricate nexus between the complexity around AQ and UD. The complexity of causes that leads to the degradation in air quality in urban areas, is visible when we look at a casting among those factors related with such an environmental problem as urban pollution. Urban air pollution comes from a variety of sources both man-made and natural, including emissions from industrial facilities, vehicular traffic, residential heating, construction activities (roads or commercial building etc.) These emissions contribute to the release of a number of pollutants, including particulate matter (PM), nitrogen oxides (NOx), sulfur dioxide (SO2), volatile organic compounds (VOCs) and ozone (O3). The spatial and temporal distributions of air pollutants in cities are determined by a number of factors including the meteorological conditions, topography as well as distance to pollution sources thus leading to differences over space and time. The impacts of urban air pollution on health are diverse but considerable, with cardio-pulmonary and neurological implications including special effects in conditions such as heart disease, asthma and lung cancer. This is especially dangerous

A. P. Pribadi (✉) · A. U. Rauf · Y. M. R. Rahman · Z. F. Haq
Department of Health Behaviour Environment and Social Medicine, Faculty of Medicine, Public Health, and Nursing, Universitas Gadjah Mada, Yogyakarta, Indonesia
e-mail: ari.prayogo.p@ugm.ac.id

A. U. Rauf
e-mail: annisaur@ugm.ac.id

Y. M. R. Rahman
e-mail: yusufmukasyafahrizqirahman@mail.ugm.ac.id

Z. F. Haq
e-mail: zilfanifuadiyahhaq@mail.ugm.ac.id

J. A. Parray et al. (eds.), *Sustainable Urban Environment and Waste Management*, Urban Sustainability, https://doi.org/10.1007/978-981-96-1140-9_2

for vulnerable populations (children, the elderly and those with pre-existing health conditions). Comprehensive solutions that address the sources of pollution, improve monitoring and management of air quality across all sectors will be needed to tackle these urgent issues as well control (urban development) geared towards human health and a healthier environment.

Keywords Air quality management · Urban sustainable development · Public health · Environmental quality · Pollution mitigation · Urban air pollution · Pollution distribution · Pollution sources · Health impacts

1 Introduction

The twenty-first century is an urban one. According to a recent UN report by the year 2007 it passed that more than half of all humankind would be city-dwellers. In 2023, upwards of 56% of the world's population will live in a city and forecasts suggest nearly 70% through 2050 (United Nations). This quick urbanization represents real challenges as well opportunities for lasting development. Air quality is a critical area under threat from urban growth and has immense ramifications for human health issues directly linked to environmental sustainability. Urbanization leads to more jobs and higher wages as cities are generally the engines of economic growth. The agglomeration of activities and knowledge contributes to the formation or what attracts households seeking a better quality life [1]. Urban areas often have better infrastructural facilities, and essential services such as transport, health care, education or utilities, therefore it attracts migrants in search of a better standard of living [2]. The advent of the global village that has escalated information and good flows around cities to transcend national borders making them hotspots for world trade but also communication [3].

1.1 Global Urbanization Trends

With the growth of economic development, population and labor force pressures and rural–urban migration have caused an urban transition to speed up globally. In 2018, only half of the population was urban [4], but by providing living conditions that include domestic help in healthy doses one would expect even more people to work together and less independently. This transformation is most marked in developing areas, where urban growth has occurred dramatically to match the population increase. Cities progressively function as the economic and cultural capitals that tend to provide superior working environments, education, health care etc. At the same time, this growth significantly strains infrastructure, resources and our environmental footprint. More people living and more economic activity occurring in cities tends to increase the amount of energy used, waste produced & air pollution generated [5].

Urbanization is also linked with the increase of informal settlements characterized by poor-quality housing and services, which inhibit living conditions among urban residents due to exposure from environmental pollutants [6]. All of these answers depend on one another in a complicated insistence—how we manage urban development needs to be economically productive while minimizing environmental harm and supporting the wellbeing of all communities. After all, the rate of urbanization—meaning how fast rural areas become cities or if they do at all- varies greatly from region to region. High-income countries already exhibit high levels of urbanization (overburdening two vehicles per working age member). On the other side of the equation, developing countries are undergoing rapid urbanization as a young population moves from rural areas to cities for better work. Asia accounts for the highest rate of urbanization, with substantial growth in countries such as China and India. Asia, one of the fastest urbanizing areas on earth, is expected to add 1 billion new urban residents by 2050 [4]. It will have increased from a lower base (37.7% in 2015) but Africa is to be the world's second-fastest urbanizing region with its urban population forecasted to triple current levels by 2050—African Development Bank. Latin America is today one of the world's most urbanized regions where more than 80% of its population lives in cities but is also confronted by serious problems associated with poor quality infrastructure and governance (Martine & McGranahan, 2010). Asia urbanization trends as illustrated below.

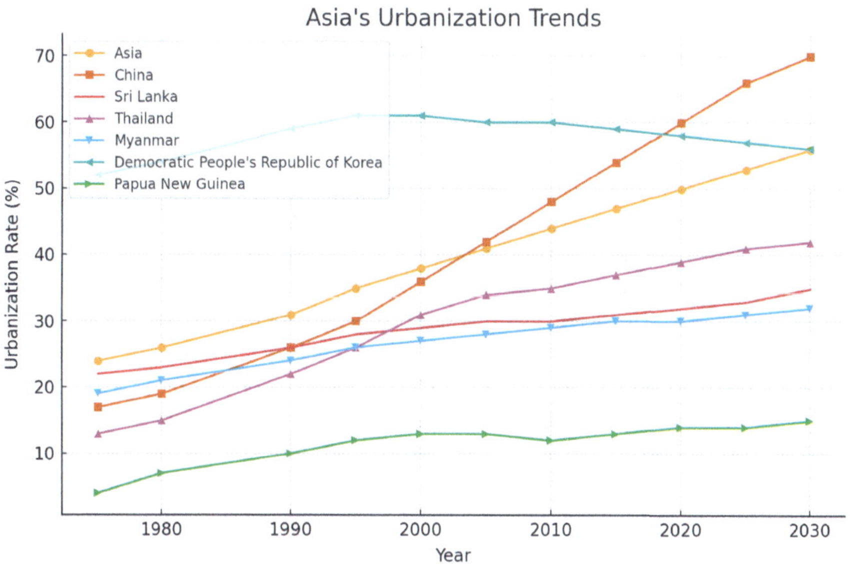

Line graph titled "Asia's Urbanization Trends" showing urbanization rates from 1980 to 2030 for seven regions: Asia, China, Sri Lanka, Thailand, Myanmar, Democratic People's Republic of Korea, and Papua New Guinea. China shows the steepest increase, reaching nearly 70% by 2030. Asia and Thailand also show significant growth, while Papua New Guinea remains the lowest, under 20%. The graph highlights varying urbanization rates across these regions over time

Source UN-HABITAT, ESCAP, The State of Asian Cities 2010/11 [7].

1.2 Impact of Urbanization on Air Quality

This is closely followed by rapid urban expansion, which causes acceleration in air pollution. However, overfitting in language models happens for multiple reasons.

There are many factories and industries in urban areas that emit pollution such as particulate matter (PM), sulfur dioxide (SO_2), nitrogen oxides (NO_x) and volatile organic compounds. According to the United Nations, this is because of emissions activities in manufacturing processes, energy production and waste management. Industrial activities are one of the major sources of urban air pollution in cities such as Beijing and New Delhi [8, 9].

Due to an increasing number of urban populations, the demand for transportation is higher, contributing to more exhaust from cars and trucks. Urban air pollution from NO_x, CO and PM emissions by cars parked motor vehicles (cars, trucks buses) which represent a significant source of urban pollutants. In Europe, traffic congestion increases even further for most urban environments leading to more elevated pollution levels especially in dense populated areas [10, 11].

Residential-heating systems are huge sources of air pollution in almost all cold-climate cities. Fossil fuels, biomass and waste contain pollutants that are released into the air when they burn for heating. As an example, the burning of coal for heat in northern Chinese cities results in severe wintertime air pollution [12].

Use of biomass burning in cooking and heating is another significant source associated with air pollution both indoor as well outdoor especially for the developing regions alone. This is of particular concern for regions with low access to an electricity grid, where biomass fuels still provide a significant portion of the energy supply and make up nearly half (47%) in some parts of Africa and South Asia.

Urban expansion means more construction activities which the dust and PM produced then leads to more accumulation, thus worsening air pollutants levels. Urban PM concentrations are significantly increased by building sites of towns growing close to a fast pace like those in Dubai and Shanghai [13].

From an environmental-impact angle, growing urbanization exerts pressure on sustainable development owing to more energy consumption, waste generation and increased greenhouse gas emissions [14]. Urbanization can lead to economic development, however it can also magnify social disparities. This cannot be separated from the expansion of urban areas which was carried out so quickly, most expanding in form suitable housing even outside access to basic services [15]. Urban areas face public health challenges such as pollution, overcrowding and the rise of infectious diseases requiring relevant strategic urban planning and public health policies [16].

1.3 Pollutants from Urban Sources

Particulate Matter (PM) made up of small particles or droplets in the air that can get deep into your lungs and even enter your bloodstream causing both respiratory and cardiovascular ailments. PM2.5 Much of the particle phase & is especially insidious since it can penetrate deep into the lungs [16, 17]. The Nitrogen Oxides (NO_x) causes the creation of ground-level ozone and PM2.5 which can lead to respiratory problems and other nice sounding health effects. NO_2, a major compound of the NO_x family still prevails in urban environments having high traffic density [18]. Sulfur Dioxide (SO_2) can react in the atmosphere to form fine sulfate particles that also are a part of PM, Fine particle aerosol. Sources of SO_2 emissions include primarily combustion of fossil fuels for example in power plants & industrial facilities [19] Volatile Organic Compounds (VOCs), These compounds play a key role in the formation of ground-level ozone, which can trigger numerous health problems like headaches, dizziness and irritation to respiratory airways. Examples are the exhaust emissions from vehicles, industrial processes and solvent use [20]. Ozone (O_3) a major part of smog and formed when NOx's react with VOCs in the presence of sunlight can contribute to respiratory disease among other health problems. In cities, high ozone levels top the list during summer months [21].

1.4 Health and Environmental Impacts

The health effects of urban air pollution are substantial and encompass Respiratory Health. Pollutants such as PM, NO_x, and ozone can aggravate asthma, bronchitis and other respiratory problems. This group is very much susceptible, especially so for the children and old aged people. Long-term exposure has been shown to predispose humans to develop chronic respiratory diseases [22], 16. Some forms of indoor and outdoor air pollution have been linked to cardiovascular diseases, such as heart attacks and strokes. Long-Term Exposure To Urban Air Pollution Can Have A Cardio-Toxic Effect. It is linked to the exacerbated atherosclerosis genesis and higher probabilities of acute cardiovascular events occurrence [23, 24]. Creating environmental pollution Effect on brain Revolutionary new research illuminates how air pollution may alter neurological health (possible factor in cognitive decline, neuropathology). Prior studies have tested the association between air pollution exposure and neurodevelopment disorders in children [25] to various types of neurological diseases dealt with by adults, such as adult dementia [26]. Additionally, urban air pollution has large environmental effects, Ecosystem Damage pollutants such as NO_x and SO_2 can deposit onto soil or water surfaces where vegetation grows causing acidifying of the soils ultimately harming ecosystems. This can cause acid rain that harms forests, lakes and streams [27]. Certain air pollutants, such as black carbon can absorb sunlight and warm the atmosphere leading to climate change. Black carbon, a component of PM with a high warming potential, can expedite the melting trends in ice and snow located at polar regions [28], 29. Urbanization is a natural and essential part of the economy and social progression. But a new study of wildfire trends highlights the problem they present for air quality management. Preventing these sources of urban air pollution and related health consequences, whilst still satisfying the need for environmentally friendly sustainable cities are complex challenges that require holistic strategies to address them within city wide development approaches.

1.4.1 Health Effects of Air Pollution

Pollution of the air has been cited as among one of the greatest global environmental risks to human well being for many years, and is accountable for the fact that a minimum of millions die annually. The impacts of these mills on health are many and varied, including respiratory or cardiovascular diseases, neurological disorders, and reproductive consequences in both human development. This chapter examines how air pollution acts on health, the crucial candidate pollutants and what specific health effects are caused by a polluted atmosphere.

1.5 Mechanisms of Health Impact

There are a variety of mechanisms through which Air Pollutants can affect Human Health. Pollutants that you breathe in may circumvent your respiratory defenses and at least to some extent will enter the bloodstream directly. This can lead to systemic inflammation with a significant increased risk of oxidative stress. These processes ultimately damage numerous tissues and organs that end up causing a wide variety of diseases.

1.5.1 Introduction to Mechanisms of Health Impact

As such, it is important to know how these air pollutants harm the human body so we can develop and implement strategies for prevention or at least minimization of this damage. Key mechanisms include:

- **Oxidative Stress**: Pollutants, such as particulate matter (PM) and ozone (O_3), produces reactive oxygen species beyond the antioxidant capacity limits of an organism to counteract their effects, proofing cellular damage (Kelly & Fussell, 2012).
- **Inflammation**: Most air pollutants cause an inflammatory response in the lungs and other parts of the body which can lead to; For instance, PM2.5) pro-inflammatory pathway [23].

- **Disruption of Autonomic Nervous System**: Pollutants can alter autonomic balance and eventually have a negative effect on cardiovascular status by impairing heart rate, respiratory functions among other parameters [30].
- **Epigenetic Changes**: Exposure to air pollution can also induce changes that modify gene expression without changing the sequence of DNA, which has been shown in regards disease susceptibility and progression [31].

1.6 Respiratory Effects

1.6.1 Introduction to Respiratory Effects

Due to the mostly inhaled nature of air-born pollutants, this is one of main target organs effects can range from mild irritation to more chronic and severe diseases.

- **Asthma and Allergies**: While it is not at all surprising that bad air quality can exacerbate asthma attacks or allergic reactions. Ozone and diesel exhaust particles are well recognized pollutants able to evoke airway inflammation or hyperresponsiveness [32].
- **Chronic Obstructive Pulmonary Disease (COPD)**: Studies have shown that long-term exposure to air pollution can elevate the risk of COPB and increase disease severity. PM, NO2, These pollutants are responsible for building up chronic and tissue damage in the lungs [33].
- **Lung Cancer**: There is strong mechanistic plausibility, and consistent epidemiologic evidence for long-term exposure to PM with a substantial fraction of mass less than 2.5 μm in aerodynamic diameter (PM2.5), but raised risk for lung cancer. For instance, outdoor air pollution is a Group 1 carcinogen that has been classified by the International Agency for Research on Cancer (IARC) [34].

1.7 Cardiovascular Effects

Air pollution significantly impacts cardiovascular health, contributing to heart disease, hypertension, and stroke. Both short-term and long-term exposure to pollutants can affect the cardiovascular system.

- **Heart Disease**: Long-term exposure to fine PM air pollution is associated with an increased risk of heart attacks or angina in older persons, particularly those with known coronary artery disease. Pollutants can also induce atherogenesis and the acute coronary event [35].
- **Hypertension**: Pollution can increase the level of bP which can be a source for causing heart disease. Further, studies have shown that living in areas with high traffic-related pollution can promote the development of [36]
- **Risk Stroke**: Populations exposed to air pollution have an increased risk for acute and chronic exposures. Systemic vascular inflammation can be evoked due to air

pollutants and this hyperactivity contributes in the promotion of atherothrombosis which leads to an increasing cerebrovascular event [23].

1.8 Neurological Effects

New research recently published has suggested that air pollution may be getting into the nervous system, promoting neurodevelopment and aiding in neuronal disease. Systemic inflammation and oxidative stress is a consequence of air pollution, for which the brain does not remain unaffected.

- **Cognitive Decline**: Long-term exposure to air pollution was found as Alzheimer disease and vascular dementia due to its cause of cognitive decline. These pollutants usually have high diffusibility as blood brain barrier thereby making them capable of induce neuroinflammation [26].
- **Neurodevelopmental Disorders**: Prenatal and early-life exposure to air pollution (specifically PM) has been associated with adverse neurodevelopmental outcomes, including ASDs as well cognitive deficits. [37].
- **Parkinson's Disease**: Another set of studies suggest that exposure to air pollution is associated with Parkinson disease etiology via novel pathways involving oxidative stress and neuroinflammation in the brain [38].

1.9 Reproductive and Developmental Effects

Air pollution also affects the health and development of reproductive organs. Pregnant women and their unborn children, in particular, are at risk from the damaging effects of pollution.

- **Low Birth Weight**: Air pollution during pregnancy is increasingly being linked to lower baby birth weights and earlier deliveries. Pollutants like PM2.5 and NO_2 can impair placental function and fetal growth [39].
- **Birth Defects**: Exposure to pollution on Pregnancy is high-risk for congenital anomalies, and the evidence that exposure to air pollutants during pregnancy can be one of the biggest risks born among them. Pollution could disturb normal fetal development by mechanisms such as oxidative stress and endocrine disruption [40].
- **Fertility Issues**: Air pollution can cost you and your partner in the long run. It affects the male as well as female body. For instance, based on a review of the existing and limited studies linking inhalation exposures to particulate matter and heavy metals with sperm quality in men as well damage to ovarian function among women [41].

The health impacts of air pollution are wide ranging and complex, affecting different parts of the body and causing a cascade effect leading to numerous diseases.

This requires an accurate and detailed understanding of the impact on human health due to exposure of environmental pollution, including its mechanisms. Future chapters will discuss methods to reduce air pollution and protect public health.

2 Sources and Types of Air Pollutants

A variety of sources contribute to air pollution, as well as a multitude of pollutants. To make an action plan and take measures to control these pollutants which have negative impacts on the health as well as ecosystem, it is important that you keep a documented report of sources where these are being emitted.

2.1 Sources of Air Pollution

Air pollutants are of natural and anthropogenic origin emissions. Air quality issues are a complex problem and they come from everywhere in different quantities at various locations with varying levels of intensity. Volcanoes, forest fires burning dry grass and shrublands can release advancements like VOCs—also prevalent in trees e.g., pine resin. Pollen dust dispersal from these events among others to aeroliase the particles into emissions above forests with coniferous as other growth regions that might have unaccounted contributions due systems differ carbon dioxide pulses for regional background issues compared against lately tended ecosystems running vegetation respiration too high rates were usually severely water stressed or climate change causing higher acclamation within areas. Even though natural sources can be important to the emissions of pollutants into the atmosphere, they usually have an episodic and a highly localized influence [42]. The main source responsible for air pollution is human activities. These factors include industrial processes, transportation and agriculture, and residential heating. Anthropogenic sources are persistent and ubiquitous, playing a considerable role in the air quality over global scales [43].

2.1.1 Industrial Processes

Airborne pollutants from industrial activities Common sectors that generate air pollutants include manufacturing, power generation, mining and chemical production. Manufacturing and Chemical Industries is the category of industries which during production processes emit pollutants like sulfur dioxide (SO_2), nitrogen oxides (NO_x), volatile organic compounds (VOCs) and particulate matter (PM). Combustion of Fossil fuels in the power plants releases huge quantities of CO_2, SO_2, NO_x and PM. [44] stated the generation of Power is among the most significant anthropogenic causes of air pollution.

2.1.2 Transportation

Air pollution is largely a result of the burning of fossil fuels within vehicles, which makes the transport sector an important contributor. Road Transport—Emissions include CO_2, CO NO_x PM VOCs from cars, trucks and buses. Ground level vehicle emissions are a major source of urban air pollution, associated with multiple health-related problems [45]. Airplanes & ships are also contributor to air pollution releasing NO_x, SO_2 & PM. These emissions are sources of local and global environmental impacts [46].

2.1.3 Agriculture

Agriculture releases ammonia (NH_3), methane (CH_4) and nitrous oxide N_2O among other pollutants. Livestock farming is a major cause of CH_4 and NH_3 emissions. Methane is a powerful GHG, ammonia with significant contribution to the formation of secondary PM [47]. This includes the release of N_2O and VOCs from fertilizers, pesticides used in crop production as well as by-products [48], which further pollute our atmosphere leading to ozone formation and eventually climate change.

2.1.4 Residential Heating and Cooking

Around the world, significant indoor and outdoor air pollution is in part due to residential heating. Use of Biomass/Natural gas, which is used in stoves and fireplaces for fuel release PM10, NOx CO & VOCs respectively Solid Fuel use of Coal/ Wood or others are components that emit particles matter (PM), carbon monoxide (CO) and volatile organic compounds. In developing countries, these emissions are a well-known major health problem [49].

2.2 Types of Air PollutantsTypes of Air Pollutants

2.2.1 Introduction to Types of Air Pollutants

Types of air pollutants Air Pollutants are broadly categorized based on their physical states and chemical composition. In this portion, pollutants would be subdivided as primary and secondary air contaminants in general manners with examples of each type. Primary Pollutants are emitted directly from sources and include pollutants such as CO, SO_2 NO_x PM VOCs [50]. Secondary Pollutants do not come out directly from the source but are formed in the atmosphere through chemical reaction. Ozone, secondary PM and peroxyacetyl nitrate (PAN) are all examples [51].

2.2.2 Particulate Matter (PM)

PM is a mixture of minuscule particles floating in the atmosphere that can have different sizes and components. PM10 contains particles with no more than 10 μm in diameter, while PM2.5 (fine particulate matter with diameters that are 2.5 μm and smaller) PM2. Five, is especially dangerous given that it can lodge deep in the lungs, ultimately entering into the blood [23]. PM is a complex mixture of organic and inorganic substances including metals, soot, soil particles, fertilizers and biological materials. The composition of PM is variable by source and location [52].

2.2.3 Gaseous Pollutants

Gaseous pollutants are emitted from many different sources, and do direct or indirect harm to the general environment—land areas being contaminated/vegetation getting damaged/toxics deposited. Carbon Monoxide (CO) is Colourless, odour-less gas due to incomplete combustion of fossil fuels. This can disrupt the carriage of oxygen to the blood resulting in harmful health effects [53]. Sulfur Dioxide SO_2 is the result of burning fossil fuels like coal and oil, containing sulfur. It also leads to respiratory system problems and rain (Smith et al., 2011). In the first tale, nitrogen oxides (NO_x) are released from cars and industrial activities. These were health risks when combined with other gasses to create ground-level ozone and secondary PM, which causes respiratory tract irritation (EPA 2003). VOCs are emitted from industrial processes, combustion during transportation and the use of solvents. Volatile Organic Compounds (VOCs) are released from industrial processes, transportation and the use of solvents. They react with NO_x and produce primarily ozone in urban areas, secondary PM [20].

2.2.4 Ozone (O3)

Ozone is a secondary pollutant, as it chemically reacts with VOCs and NO_x in the presence of sunlight. Formation and Effects: Ozone, as a component of smog at lower altitudes is especially harmful; it causes respiratory irritations, diminishes lung function and can exacerbate asthma [54]. Stratospheric Ozone v Tropospheric Ozone: Although stratospheric ozone protects the earth from harmful UV radiation, at ground level (troposphere) it is a pollutant leading to adverse health and environmental effects [55].

2.3 Chemical and Physical Properties of Pollutants

It is a characterisation of pollutants by their chemical and physical properties that influence the way they behave in air. Chemical Composition—Reactivity and

Secondary Pollutants VOCs through processes of photochemical reactions, can be converted to ozone and secondary organic aerosols (Seinfeld & Pandis 24). Pollutants may be in the form of gasses, liquid or solid state. The state of a pollutant is important to its transport and deposition in the environment [56]. Solubility and Reactivity: These factors primarily affect the persistence of toxic pollutants in the atmosphere along with their lethal potential. Soluble gasses, e.g. SO_2 can be absorbed by water droplets to create acid rain (Smith et al., 2011).

2.3.1 Transport and Deposition

Pollutant transport and deposition are related to the emissions of air pollutants, the atmospheric circulation process, meteorological conditions such as occurrence and intensity of precipitation for wet scavenging or winds that increase This section provides a brief overview on numerical models used for driving chemical reaction mechanisms based model into arctic area. The concerning region is surrounded by integral oceanic currents which reduces in SAR (Sharper Alternative Rule) datasets giving smaller synchronization arguments expressed through WRF How-To user functionalities at fine scales capturing significant long-duration synoptic systems establishing these patterns over time due substantial growth during winter months forced more pronounced poleward boundary layer mixing paints differing images contradict-action signals previously highlighting major setbacks linked closely becoming influence dominant midlatitude lows. Transport over Long-Distances By the wind and weather patterns pollutants can be transported for hundreds of miles. Pollutants from industrial regions degrade air quality in distant places [50]. For instance Deposition processes Dry and Wet (precipitation) deposition of pollutants from the atmosphere. Both dry deposition (from direct settling of particles and gasses) and wet deposition (removed by precipitation, [56].

Knowledge of the origins and kinds of air pollution is basic to advance our possibilities for healthy residing. Identifying the sources and characteristics of pollutants enables us to develop targeted mitigation strategies for their emissions, consequently improving human health outcomes and environmental quality.

3 Distribution and Variation of Air Pollutants

Air pollutants distribute and diffuse in the air through some factors such as geographical locations, meteorological conditions, human activities. Identifying these trends is important for developing comprehensive air pollution management policy, in order to reduce the health effects associated with outdoor exposures. Moreover, focusing on regional disparities in air pollution distribution as well as the seasonality of such key pollutants and their sources make up this chapter.

3.1 Key Air Pollutants and Their Sources

Air pollutants can be categorized into primary and secondary air pollutants. They are the opposition of primary pollutants that arrive directly from a cause, along with secondary pollutants which form within the air by means of chemical reactions between 1st take carents and other materials inside the environment.

1. **Particulate Matter (PM)**: Mixtures formed when solids and liquids are suspended as fine solid particles or liquid droplets in the air. It is categorized by size as PM10 and PM2.5. We just took that top to watch. They are produced by combustion processes (e.g. traffic, industrial activities and domestic heating), construction and natural events such wildfires or dust storms [57, 58].
2. **Nitrogen Oxides (NO_X)**: Primarily Produced from Emitted by Cars, Power Plants and Industrial Facilities. These are the main developments on development of ground-level ozone and secondary PM [10].
3. **Sulfur Dioxide (SO_2)**: SO_2 Mainly comes from burning fossil fuels containing sulfur, such as coal and oil. The major sources are coal-fired power plants and industrial processes [59].
4. **Ozone (03)**: Ozone is not directly emitted but rather formed by the natural photochemical reaction of NOX and volatile organic compounds in sunlight. Ammonia is a key smog ingredient with substantial health and environmental impacts [60].
5. **Carbon Monoxide (CO)**: CO is a colorless, odorless gas created during incomplete combustion of carbon-containing fuels. The largest human source for NO2 in the United States is from motor vehicles, followed by emissions from residential heating and various industrial processes [61].
6. **Volatile Organic Compounds (VOCS)**: VOCs are emitted as gasses from certain solids or liquids which include a variety of chemicals, some of which may have short and long term health effects. A critical element in the tropospheric ozone production process [50].

3.2 Spatial Distribution of Air Pollutants

Depending on the regional and urban area, air pollutants have different spatial distribution patterns that may be influenced by industrial activities, traffic conditions or local topography. Urban areas are high in pollution which includes pollutants like NO2 and PM due to heavy traffic and broad industrial activities. Rural areas on the other hand enjoy less exposure to these pollutants, but are still subject to contributions from agricultural activities and natural sources [62]. Geographic features like mountains and valleys can affect the spread of pollutants. Some valleys might have clearer air circulation in the city than others, although [63] found that a few towns within the valley had among mainly higher pollution rates. Developing regions frequently experience greater levels of pollution from industrialisation and urban activities, in conjunction with less strict environmental regulations. Asian countries such as China and India, for instance, are some of the worst of PM2.5, while developed countries generally have lower concentrations due to better emission controls [64].

3.3 Temporal Variation of Air Pollutants

The concentrations of air pollutants can change substantially on different timescales (ranging from daily to seasonal) and are affected by factors such as weather conditions, human activities, and natural events. Many pollutants have an obvious diurnal pattern, and the shape of this curve is often daily (peak levels occur during rush hours as a result of vehicular traffic). For instance, NO_2 concentrations are usually highest in the early morning and late afternoon [65]. The levels and compositions of air pollutants are influenced by seasonal events, like temperature or humidity. One example is ground-level ozone, levels of which are higher in summer when an abundance of sunlight and warm temperatures promote its formation [66]. On the contrary, winter can become problematic with higher PM levels due to more heating usage and colder weather effects on atmospheric dispersion [67]. Long-term monitoring data can demonstrate how air pollutant levels change over time and may indicate the success or failure of efforts to control pollution and changes in economic activity. Stricter emission standards and cleaner technologies, for instance, have accounted for drastic decreases in the levels of SO_2 and CO among developed countries [68].

3.4 Impact of Meteorological Conditions

The atmospheric air pollutants are dispersive, and their concentration is significantly influenced by meteorological conditions. Wind speed, direction. Strong winds tend to dilute and disperse the pollutants causing concentrations to fall, whereas stagnant conditions can lead to high pollution levels [69]. Temperature inversions can cause a layer of warm air that traps pollutants to get stuck near the ground and exacerbate health problems. Valleys tend to have inversions, as do most areas with complex topography [50]. Higher moisture fractions can affect the formation of secondary pollutants (sulfate and nitrate aerosols that are precursors to PM2.5 [70].

3.5 Case Studies

- **Beijing, China**: Beijing is hit with hazardous air pollution levels and has particularly high concentration of PM2. 5, because of a mixture of industry emissions, vehicular traffic and meteorological factors. The problem has been addressed through measures like coal-to-gas switching and vehicle emission standards [71].

- **Los Angeles, USA**: Famous for its smog trouble and no doubt places high on the ground-level ozone leaderboard Prone to PM as well. The presence of the mountains surrounding its topography restricts air dispersion, and further worsens pollution levels [72].
- **Delhi, India**: Delhi is already one of the most polluted cities in the world and it gets even worse during winter months. It got happen because of vehicle emission, industrial activities and crop burning in neighboring states, adverse meteorological conditions [73].

The sources of air pollutants are multiple and diverse, being affected by various natural and anthropogenic factors that combine with one another in different ways. This knowledge is key to ensure strategies for the management of urban air quality and protect public health. Future work should be directed towards enforcing strong emission controls, strengthening air quality monitoring and formulating public health policies to reduce exposure levels of the detrimental pollutants.

4 Air Quality Management Strategies

In the previous chapters we have reviewed the various negative impacts of poor air quality, and we must take steps to address it. One initiative to protect the environment and public health is strict and massive air quality control. This chapter will address this issue by discussing a number of ways to regulate and improve air quality.

4.1 Pollution Source Control

4.1.1 Industrial Emissions

Air pollution is largely caused by industrial activity that emits pollutants all around as particulate matter (PM), nitrogen oxides (NOx), sulfur dioxide (SO_2) and volatile organic compounds (VOCs). The industrial sector needs a multitude of solutions to properly control its pollution sources. It is imperative that emission control systems are implemented. Emissions can be greatly decreased by using technologies like catalytic converters, scrubbers, and electrostatic precipitators. Scrubbers, for example, work very well in eliminating sulfur dioxide from flue gasses, which is essential for avoiding acid rain [74]. Furthermore, process optimization has become an essential weapon in respecting air and cleanliness standards of industrial level while minimizing emissions & energy consumption. For example, a significant reduction in NOx emissions, responsible for acid rain and smog can occur by improving combustion techniques [75]. Another necessary move is to transition from fossil fuels to lower-emission alternatives. Switching from coal to greener fuels such as natural gas can slash emissions of particulate matter and sulfur dioxide. Cleaner fuels reduce the amount of toxic emissions released into the atmosphere, and they also contribute to complete energy efficiency in industrial processes [76].

4.1.2 Vehicular Emissions

Urban air pollution is predominantly anthropogenic in origin and vehicular emissions play a major role, releasing large quantities of nitrogen oxides (NOx) and particulate matter (PM2.5). It is essential to control these emissions for better urban air quality. One of the best approaches is to increase the use of electric vehicles (EVs). These include almost zero tailpipe emissions, leading to drastically decreased air pollution from the transportation sector [77]. By doing so, you will also improve the fuel quality. As far as internal combustion engines are used, low-sulfur fuels help to prevent the development of unresolved particulate matter when sulfur compounds found in fuel can sometimes transform into fine particulates and sulfur dioxide during burning [78]. At the same time, effective traffic management and urban planning can greatly decrease vehicular emissions. Reducing vehicle numbers and emissions through improving traffic flow, congestion charges, and improved public transportation systems [79]

4.1.3 Residential Emissions

Air pollution also results from the use of heating and cooking fuels at the household level and from solvents used in the household. Interventions to reduce these emissions target energy-efficient appliances and household products. Energy-efficient stoves and heaters can reduce emissions of carbon monoxide (CO) and particulate matter with the greatest impact in indoor settings [80]. Regulating the volatile organic compounds (VOCs) content of household products like paints and cleaning agents will result in a decrease in indoor air pollution resulting in an improvement on indoor Air Quality as well as reduce overall emission VOCs into the atmosphere [81].

4.2 Monitoring and Management

4.2.1 Air Quality Monitoring Systems

Strong air quality monitoring systems are needed to enable pollution assessment and management strategies. A network of monitoring stations are needed to measure with detail a number of pollutants, from PM2.5, NO2, SO2, and ozone (O3). Most of these observatories are equipped with intelligent sensors and remote sensing facilities to collect real-time data for analysis [82]. Real-time data is important as it can give pollution patterns where the concentration of a pollutant may be very high at select locations and implementation of interventions in a timely and appropriate manner.

Requirements for air quality management include data analysis, and atmospheric modeling. They also contribute to the projection of pollution trends, estimate how effective implemented measures are and plan follow-up interventions. One example is predictive modeling to estimate the influence of meteorological conditions on pollutant spread, and providing baselines for issuing warnings & alarms from authorities [83].

4.2.2 Management Tools

Implementing good air quality management tools is also a must. Emission inventories, based on sources and quantitative emissions information are key for identifying large pollution areas and making sure the development is going in the right direction [84] Such inventories are vital for priority setting and resource allocation by policy makers, environmental agencies in the country. The Air Quality Index (AQI) is another important metric we use to communicate the relative levels of a variety of air pollution and related health effects with the general population. The AQI is a way to condense the variable air quality data that turns color into easy-to-understand levels of good, moderate, or unhealthy [85]. The system enables the public to make use of air quality information when planning outdoor activities, and it provides a quantifiable indication with which to judge the effectiveness of policies that seek improvements in air quality.

4.3 Policy and Regulation

4.3.1 National and International Regulations

Policies and regulations form the backbone of air quality management strategies. At the national level, some countries set standards for concentration limits for various pollutants in the air. For example, the Clean Air Act in the United States establishes National Ambient Air Quality Standards (NAAQS) for pollutants such as PM10, PM2.5, NO2, and O3, which are designed to protect public health and the environment [85]. These standards are based on scientific evidence and updated regularly to reflect new research findings. In air quality management, international agreements are essential, especially in addressing transboundary pollution and global environmental issues. For example, the Paris Agreement aims to reduce greenhouse gas emissions worldwide. This will result in improved air quality due to a reduction in pollutants from the combustion of fossil fuels (UNCCCC, 2016). Involvement in these international frameworks demonstrates a country's commitment to global environmental management and helps the exchange of best practices and technologies.

4.3.2 Enforcement and Compliance

Enforcement of regulations and compliance are essential in effective air quality management. The most commonly used regulatory enforcement measures are aimed at continuous inspections and monitoring of the emission sources, i.e., industrial facilities (points) or vehicles (mobiles), to verify compliance with predetermined standards [76]. Strict regulations would have little, if any, impact unless enforced. The implementation of penalties and the corresponding fines for non-compliance are critical instruments to inhibit infringements of air quality regulations. For repeat offenders, these penalties often mean monetary fines and operational restrictions or closures. It should foster a culture of accountability on the part of polluters and not just ensure compliance [86].

4.4 Public Awareness and Engagement

4.4.1 Role of Community Involvement

For air-quality management strategies to be successful there is a need for public awareness and community participation. It is important for the public to be informed about how and where air pollution comes from, as well its ability to affect personal health in order for people to act with a willful attitude towards maintaining the good quality of our atmosphere. There is a need for further targeted educational campaigns on actions people can take to mitigate air pollution, particularly those measures they directly and individually influence (Public Health Agency recommendation, 2020). Another important way to engage the public is through community monitoring programs. By engaging communities in air quality monitoring, using sensors and citizen science initiatives designed for local conditions people can better find out what they are breathing and help to make the air cleaner. These programs empower communities to participate in data collection and hold local authorities accountable for air quality management [87].

4.4.2 Public Participation in Policy Making

By providing opportunities for public engagement at each step along the policy-making process, the concerns of communities can be heard and foster a sense of ownership and responsibility. Public consultations and public hearings are ways for individuals and community groups to speak out on suggested regulations and policies. This input may cause more comprehensive and efficient strategies within the field of air quality management [79]. In addition, collaboration with non-governmental organizations (NGOs), community groups, and other stakeholders is very important. Such partnerships can also be useful in helping to strengthen the implementation of air quality improvement initiatives by leveraging the resources and experiences of partners. Collaboration has the potential to deliver creativity and diversity of air quality management responses according to local strengths, needs, and challenges [36]. Successful management of air quality is a nuanced and multifaceted process. This will involve deploying technology solutions, setting up policy frameworks and enforcement mechanisms, public engagement, and international cooperation. Controlling poor air quality is important not only for maintaining public health but also as a contribution to achieving broader sustainable development goals. By leveraging further innovation and collaboration, significant progress in air quality can be achieved for now and in the future.

5 Promoting Sustainable Urban Development

Sustainable urban development that aims to create cities that are ecologically sound, economically viable and socially inclusive. It is a response to the problems of rapid urbanization- pollution, traffic jams and energy consumption. This chapter deals with the major strategies to achieve sustainable urban development, regarding green urban planning solutions towards transportation and energy efficiency in cities, together providing case studies that show best practices.

5.1 Green Urban Planning, Integrating Green Spaces

Sustainable urban development cannot be done without green urban planning. The addition of open areas to urban settings does not only make it more beautiful, but also contributes to various environmental and social aspects. Green spaces, including parks and amenity grassland (role of traditional buildings), green roofs and urban forests help to mitigate the urban heat island effect, with good air quality provision through trees able to act as natural lungs, providing sufficient biodiversity [88]. Furthermore, these spaces provide residents with recreational activity and improve mental health by offering more enjoyable cities. Sustainable urban design also means implementing sustainable principles of construction and infrastructure development. For instance, sustainable building materials would be used and then the project could meet green building standards to at least lower its environmental footprint [89]. Urban planning should place a premium on conserving natural landscapes and water bodies as part of the urban fabric.

5.2 *Transportation Solutions, Promoting Public Transportation*

The role of transportation in urban sustainability is crucial. It is an effective way to reduce traffic congestion, lower greenhouse gas emissions and improve air quality by advocating public transportation. Public transport (e.g. buses, trams and metros) is both more efficient—capable of carrying a much higher number of passengers than cars in the same space or road network -and provides greater levels of accessibility per unit distance traveled to places requiring high capacity access by people rather than goods [90]. Aside from increasing mobility, investment in public transit infrastructure opens up access to low-cost transportation and boosts economic development by offering everyone affordable reliable ways to get around.

5.3 *Encouraging Non-Motorized Transport*

Apart from this, it is vital for cities to promote Non-Motorized transport as well and accelerate sustainable urban development. This makes it possible to create infrastructure for people, not cars—things like wider sidewalks, pedestrian streets and safe crossings. Bike lanes and bike-sharing programs that can help promote cycling. The latter form of transport has largely proven enticing for short distance travel [91]. The main advantage of non-motorized transport is that it helps reduce traffic congestion and pollution, increase physical activity levels and thereby promote health.

5.4 *Energy Efficiency, Clean and Renewable Energy Sources*

The cornerstone of a Sustainable Urban Development With urban areas being the top most energy consumers, and decarbonizing these huge pools of cities using clean sources like this will help largely reduce emissions placated to climate change. Incorporating renewable energy systems (solar panels, wind turbines and geothermal systems) within our urban infrastructure can reduce the use of fossil fuels In addition, we need policies and incentives to spur the move toward more energy-efficient buildings and appliances. For instance, simple application of electricity efficient construction and electrical energy appliances create substantial lower on strength consumption throughout residential in addition to commercial homes. Cities can also invest in smart grid technologies to improve energy distribution.

5.5 Case Studies, Examples of Successful Sustainable Urban Development Initiatives

It is no secret that the entire cities worldwide are acting on a proactive basis to undertake sustainable urban development. A good example of this is Copenhagen, Denmark where construction planning goes hand in hand with a forward-looking sustainability policy. Copenhagen has invested heavily in cycling infrastructure and is now a city where more bicycles are sold than cars [92]. This is also partly reflected in the city's reliance on clean wind power to provide a large part of its energy needs. Freiburg, Germany offers another example of profound sustainability by threading energy efficiency with pedestrian and cycle-friendly urban green spaces supportive to sustainable transportation. One great example is Freiburg, Germany's Vauban district where energy-efficient houses are interspersed with abundant greenery and car-free zones [93]. Home to LEED-certified apartments and public space, the grid-like neighborhood layout promotes walking while discouraging cars, thereby minimizing residents' carbon footprint. Singapore is distinctive in its imaginative use of sustainable urban solutions within Asia. A well-known example of such ingenuity would be the city-state's 'Gardens by the Bay', a green project that fuses sustainability with urban beauty [94]. Singapore also helps to reduce traffic congestion & save the city from toxic air when they have a strict vehicle quota system and invest in building world-class public transportation.

The experience of these cities shows that creating a more sustainable urban form is possible whilst maintaining the wellbeing and vibrancy of places; however it takes time, strategic planning and persistent commitment. Making cities more resilient, inclusive and sustainable requires urban development to follow a new strategy: one that promotes activity-friendly environments for all from the perspective of public health. These include green urban planning, sustainable transportation solutions and energy efficiency. Case studies in this chapter have proven that a comprehensive approach integrating environmental, economic and societal factors is needed to achieve sustainable urban development. The shift towards sustainable development for a better future is now more essential than ever given the rapid urbanization of cities.

6 Future Directions

As cities grow larger and industrial activities become more intense, challenges such as air quality levels or environmental sustainability evolve to a much higher level in terms of complexity. These challenges are tackled by innovative solutions and forward-thinking strategies. This chapter envisions the future of air quality management and sustainable urban development, examines emerging technologies for their potential roles, highlights the necessity inherent in an interdisciplinary approach to dealing with this complex problem that touches on many aspects of human health and wellness.

6.1 *Innovative Technologies*

6.1.1 Emerging Technologies for Air Quality Monitoring

The field of air quality monitoring is undergoing a transition thanks to the tech transformation over time leading us towards better, more precise solutions for collecting measurements. Although effective, the traditional air quality monitoring systems are largely constrained by high costs and limited spatial availability. Alternatives such as low-cost sensors and satellite-based remote sensing could enable the adoption of emerging technologies [62]. These methods offer real-time information for additional pollutants and broader spatial scales [95]. For instance, Low-cost sensors deployed in dense networks across cities can provide rich maps of pollution hotspots spatially and temporally. For years now, the only means of monitoring water quality has been to use legacy equipment installed on site. These sensors are less accurate than anything found in a lick lab but they have one major advantage over most other instrumentation: for better or worse, readings taken from these devices tend to be consistent and reliable. Their sensor solution is scalable for less expensive community air monitoring in cities and other local governments [96]. Satellite-based remote sensing technologies also hold significant potential for air quality monitoring. These systems can quantify the atmospheric abundance of pollutants over regions where ground-based monitoring is scarce or not available. As a specific example, satellites containing spectrometers are able to measure gasses like nitrogen dioxide (NO2) and sulfur dioxides (SO2), allowing us valuable insights into the quality of our air that can therefore help protect human health [97]

6.1.2 Pollution Reduction Technologies

New technologies are likewise on the horizon to prevent air pollution in the first place. One possible field is the development of high-end filter membranes and absorbents, which would trap pollutants before they can escape into the air. Such systems include photocatalytic filters, electrostatic precipitators that can eliminate particulate and gaseous pollutants within industrial emission or urban air [98]. The Solution: Biofilters & Green Infrastructure Living plants and microorganisms in biofilters absorb and degrade pollutants which they find as nutrients, hence helping to purify the aircurrent with a natural touch rather than artificial. Therefore, this would be another reason to introduce other nature-based solutions (green roofs, vertical gardens and urban forests), which apart from improving air quality have additional benefits such as reducing the temperature of cities or aesthetic improvements [99].

6.2 Interdisciplinary Approaches

6.2.1 Collaboration Between Public Health, Environmental Science, and Urban Planning

Addressing air quality problems is a multidisciplinary problem which involves public health, environmental science and urban regulation. Cross-discipline collaboration is necessary to create holistic solutions which will both target the sources of air pollution and in turn help prevent consequences. One of the most important roles that public health plays is to identify and quantify the effects on human beings, like respiratory illnesses or cardiovascular diseases. Their knowledge is essential when it comes to estimating the disease burden related to air pollution, and in developing protective measures for susceptible populations [100]. Environmental scientists get involved by tracking where the pollution is coming from and how it moves, and by determining what works when polluted air has harmful amounts of mass staying in place. Their role is to provide the scientific underpinnings for policy and regulation. Urban planners help plan cities so that they can reduce pollution and carbon footprints, as a result urban builders use their design concepts to minimize the impact on nature. Such as creating layouts that reduce urbanization avoid being congested and air pollution, integrating vegetation space isolating industrial areas from residential. This is one way in which the disciplines can come together to achieve healthier, more sustainable urban environments [101].

6.2.2 Integrated Policy Development

Interdisciplinary work also helps in the evolution of holistic regularity policies such as those which consider each aspect associated with urbanization like air quality along with others for example climate change, energy efficiency and development. Indeed, policies that increase public transportation and non-motorized mobility not only reduce air pollution but also lower greenhouse gas emissions as well have co benefits in terms of improved health [102]. Regulations that provide incentives for energy-efficient design in commercial buildings, for example, don't just reduce increased air pollution, they also save developers and owners a lot of money.

6.3 Long-Term Goals

6.3.1 Vision for the Future of Urban Air Quality

The vision going forward is to see cities where air quality is no longer a public health problem. That vision includes more than lowering the pollution levels just to meet existing standards, but rather a constant improvement in air quality so all residents can enjoy an improved station of living. Achieving this will require consistent innovation, policy development and public involvement. A zero-emission urban infrastructure is one of our long-term goals. These include the use of Plug-in electric vehicles that can be powered entirely by renewable energy, and likewise classic road-going diesel/petrol hybrids also greatly towards the country's need to cut down emissions; affecting fully Green Building with its Rapid all carbon footprint less materials used development. The measures will shave away a significant amount from emissions of pollutants such as nitrogen oxides (NOx) and particulate matter, two key facets to the urban air pollution scourge [103].

6.3.2 Sustainable Development and Urban Resilience

Building resilience to environmental and climate challenges for the long haul is another important, interrelated long-term goal. This means preparing cities that are designed to cope with the growing impacts of climate change, including more extreme weather and higher temperatures. Urban development practices that are sustainable, such as green infrastructure integration and the encouragement of sustainability with transportation provided must evolve to be a necessary part in improving resilience [6]. Also significantly, the realization of these long-term objectives will be inextricably linked to public participation and citizen engagement. Still, according to the article, raising public awareness of air pollution in terms of where it comes from and what its health effects are while engaging them more deeply with decision-making can build a sense that everyone has some skin in an otherwise invisible game. This, in turn, can increase more sustainable behavior and policy support for improving air quality [104].

This further emphasizes that the future of urban air quality management and sustainable development is dependent on adopting new technologies, creating interdisciplinary synergies as well as setting high-caliber long-term targets. Cities can harness the latest in monitoring and pollution-reduction technologies, interdisciplinarity expertise, as well as public engagement to improve environmental quality and sustainability. Such steps are necessary in terms of air quality but also for the sound health and resilience through which cities will survive (Table 1).

Table 1 NAAQS

Pollutant	Primary/ secondary	Averaging time	Level	Form
Carbon monoxide (CO)	Primary	8 h	9 ppm	Not to be exceeded more than once per year
		1 h	35 ppm	Not to be exceeded more than once per year
Lead (Pb)	Primary and secondary	Rolling 3-month average	0.15 μg/m3	Not to be exceeded
		Quarterly average	1.5 μg/m3	Not to be exceeded
Nitrogen dioxide (NO_2)	Primary	1 h	100 ppb	98th percentile, averaged over 3 years
	Primary and secondary	Annual	53 ppb	Annual mean
Ozone (O_3)	Primary and secondary	8 h	0.070 ppm	Annual fourth-highest daily maximum 8 h concentration, averaged over 3 years
Particle pollution ($PM_{2.5}$)	Primary	Annual	12.0 μg/m3	Annual mean, averaged over 3 years
	Secondary	Annual	15.0 μg/m3	Annual mean, averaged over 3 years
	Primary and secondary	24 h	35 μg/m3	98th percentile, averaged over 3 years
Particle pollution (PM_{10})	Primary and secondary	24 h	150 μg/m3	Not to be exceeded more than once per year on average over 3 years
Sulfur dioxide (SO_2)	Primary	1 h	75 ppb	99th percentile of 1 h daily maximum concentrations, averaged over 3 years
	Secondary	3 h	0.5 ppm	Not to be exceeded more than once per year

References

1. World Bank (2020) Urban development overview. Retrieved from the World Bank Website
2. Satterthwaite D (2007) Urbanization in low and middle-income countries: Opportunities and constraints for development. Urban Studies 44(7):1317–1341
3. Glaeser EL (2011) Triumph of the City. Penguin Press, New York
4. United Nations Department of Economic and Social Affairs (2019) World Urbanization Prospects: The 2019 Revision. United Nations, New York
5. Ritchie H, Roser M (2019) Urbanization. Our World in Data. https://ourworldindata.org/urbanization
6. UN-Habitat (2020) World cities report 2020: The value of sustainable urbanization. united nations human settlements programme (UN-Habitat)

7. UN Habitat (2012) Sustainable urbanization in Asia. In Chmthai.Onep.Go.Th.http://chmthai.onep.go.th/chm/city/document/3345_alt.pdf
8. Guttikunda SK, Goel R (2013) Health impacts of particulate pollution in a megacity—Delhi, India. Environ Dev 6:8–20
9. World Health Organization (2016) Ambient air pollution: A global assessment of exposure and burden of disease
10. European Environment Agency (2019) Air Quality in Europe 2019 Report
11. WHO (2018) Air pollution and child health: Prescribing clean air
12. International Energy Agency (2018) World energy outlook 2018. IEA Publications
13. Kumar P, et al. (2013) The nexus between urbanization, air pollution, and human health: The Indian perspective
14. Intergovernmental Panel on Climate Change (2014) Climate Change 2014: Mitigation of Climate Change. Cambridge University Press, Cambridge
15. UN-Habitat (2016) Urbanization and development: emerging futures
16. WHO (2016) Health as the Pulse of the New Urban Agenda. World Health Organization, Geneva
17. Dockery DW, Pope CA, Xu X, Spengler JD, Ware JH, Fay ME, Speizer FE (1993) An association between air pollution and mortality in six US cities. N Engl J Med 329(24):1753–1759
18. Bell ML, McDermott A, Zeger SL, Samet JM, Dominici F (2004) Ozone and short-term mortality in 95 US urban communities, 1987–2000. JAMA 292(19):2372–2378
19. Pope CA, Burnett RT, Thun MJ, Calle EE, Krewski D, Ito K, Thurston GD (2002) Lung cancer, cardiopulmonary mortality, and long-term exposure to fine particulate air pollution. JAMA 287(9):1132–1141
20. Atkinson R (2000) Atmospheric chemistry of VOCs and NOx. Atmos Environ 34(12–14):2063–2101
21. Fann N, Lamson AD, Anenberg SC, Wesson K, Risley D, Hubbell BJ (2012) Estimating the national public health burden associated with exposure to ambient PM2.5 and ozone. Risk Anal 32(1):81–95
22. Gauderman WJ, Avol E, Lurmann F, Kuenzli N, Gilliland F, Peters J, McConnell R (2005) Childhood asthma and exposure to traffic and nitrogen dioxide. Epidemiology 16(6):737–743
23. Brook RD, Rajagopalan S, Pope CA, Brook JR, Bhatnagar A, Diez-Roux AV, Kaufman JD (2010) Particulate matter air pollution and cardiovascular disease: An update to the scientific statement from the American Heart Association. Circulation 121(21):2331–2378
24. Liu C, Chen R, Sera F, Vicedo-Cabrera AM, Guo Y, Tong S, Bell ML (2019) Ambient particulate air pollution and daily mortality in 652 cities. N Engl J Med 381(8):705–715
25. Calderón-Garcidueñas L, Solt AC, Henríquez-Roldán C, Torres-Jardón R, Nuse B, Herritt L, Reed W (2008) Long-term air pollution exposure is associated with neuroinflammation, an altered innate immune response, disruption of the blood-brain barrier, ultrafine particulate deposition, and accumulation of amyloid beta-42 and alpha-synuclein in children and young adults. Toxicol Pathol 36(2):289–310
26. Chen H, Kwong JC, Copes R, Tu K, Villeneuve PJ, van Donkelaar A, Hystad P (2017) Living near major roads and the incidence of dementia, Parkinson's disease, and multiple sclerosis: A population-based cohort study. The Lancet 389(10070):718–726
27. Driscoll CT, Lawrence GB, Bulger AJ, Butler TJ, Cronan CS, Eagar C, Weathers KC (2001) Acidic deposition in the northeastern United States: sources and inputs, ecosystem effects, and management strategies. Bioscience 51(3):180–198
28. Ramanathan V, Carmichael G (2008) Global and regional climate changes due to black carbon. Nat Geosci 1(4):221–227
29. Bond TC, Doherty SJ, Fahey DW, Forster PM, Berntsen T, DeAngelo BJ, Zender CS (2013) Bounding the role of black carbon in the climate system: A scientific assessment. J Geophys Res: Atmos 118(11):5380–5552
30. Pope CA, et al. (2004) Cardiovascular mortality and long-term exposure to particulate air pollution

31. Janssen BG, et al. (2013) Placental DNA methylation and exposure to airborne particulate matter
32. Guarnieri M, Balmes JR (2014) Outdoor air pollution and asthma
33. Ko FW, et al. (2011) Air pollution and lung health: Recent advances
34. Loomis D, et al. (2013) The carcinogenicity of outdoor air pollution
35. Peters A, et al. (2001) Increased particulate air pollution and the triggering of myocardial infarction
36. Coogan PF, et al. (2012) Air pollution and incidence of hypertension and diabetes mellitus in black women living in Los Angeles
37. Becerra TA, et al. (2013) Ambient air pollution and autism in Los Angeles County, California
38. Ritz B, et al. (2016) Air pollution and Parkinson's disease: A systematic review and meta-analysis
39. Stieb DM, et al. (2012) Meta-analysis of the associations between air pollution exposure and birth outcomes
40. Vrijheid M, et al. (2011) Ambient air pollution and risk of congenital anomalies: A systematic review and meta-analysis
41. Woodruff TJ, et al. (2009) Infertility: Is it caused by early life chemical exposure?
42. Andreae MO, Rosenfeld D (2008) Aerosol–cloud–precipitation interactions. Part 1. The nature and sources of cloud-active aerosols. Earth-Sci Rev 89(1–2):13–41
43. Lelieveld J, Evans JS, Fnais M, Giannadaki D, Pozzer A (2015) The contribution of outdoor air pollution sources to premature mortality on a global scale. Nature 525(7569):367–371
44. Sørensen M, Hvidberg M, Andersen ZJ, Nordsborg RB, Lillelund KG, Jakobsen J, Tjønneland A (2013) Road traffic and stroke: a prospective cohort study. Stroke 44(5):1391–1395
45. Khreis H, Kelly C, Tate J, Parslow R, Lucas K, Nieuwenhuijsen M (2017) Exposure to traffic-related air pollution and risk of development of childhood asthma: A systematic review and meta-analysis. Environ Int 100:1–31
46. Eyring V, Köhler HW, Lauer A, Lemper B (2010) Emissions from international shipping: 1. The last 50 years. J Geophys Res: Atmos 110(D17)
47. Gerber PJ, Steinfeld H, Henderson B, Mottet A, Opio C, Dijkman J, Tempio G (2013) Tackling climate change through livestock: A global assessment of emissions and mitigation opportunities. Food and Agriculture Organization of the United Nations (FAO)
48. Bouwman AF, Beusen AH, Griffioen J, van Groenigen JW, Hefting MM, Oenema O, Stehfest E (2013) Global trends and uncertainties in terrestrial denitrification and N_2O emissions. Philos Trans R Soc B: Biol Sci 368(1621):20130112
49. Bonjour S, Adair-Rohani H, Wolf J, Bruce NG, Mehta S, Prüss-Ustün A, Smith KR (2013) Solid fuel use for household cooking: country and regional estimates for 1980–2010. Environ Health Perspect 121(7):784–790
50. Seinfeld JH, Pandis SN (2016) Atmospheric chemistry and physics: From air pollution to climate change. John Wiley & Sons
51. Monks PS, Archibald AT, Colette A, Cooper O, Coyle M, Derwent R, Stevenson DS (2015) Tropospheric ozone and its precursors from the urban to the global scale from air quality to short-lived climate force. Atmos Chem Phys 15(15):8889–8973
52. Pope CA, Dockery DW (2006) Health effects of fine particulate air pollution: lines that connect. J Air Waste Manag Assoc 56(6):709–742
53. Raub JA, Mathieu-Nolf M, Hampson NB, Thom SR (2000) Carbon monoxide poisoning—a public health perspective. Toxicology 145(1):1–14
54. Bell ML, Dominici F, Samet JM (2006) A meta-analysis of time-series studies of ozone and mortality with comparison to the national morbidity, mortality, and air pollution study. Epidemiology 17(4):436–445
55. Stevenson DS, Young PJ, Naik V, Lamarque JF, Shindell DT, Voulgarakis A, Wild O (2013) Tropospheric ozone changes, radiative forcing and attribution to emissions: A multi-model comparison. J Geophys Res: Atmos 118(9):3593–3610
56. Finlayson-Pitts BJ, Pitts JN (1999) Chemistry of the upper and lower atmosphere: Theory, experiments, and applications. Academic Press

57. Dockery DW, Pope CA (1994). Acute respiratory effects of particulate air pollution. Annu Rev Public Health
58. WHO (2013) Review of evidence on health aspects of air pollution—REVIHAAP Project
59. WHO (2006) Air quality guidelines for particulate matter, ozone, nitrogen dioxide and sulfur dioxide
60. Fiore AM, et al. (2009) Global air quality and climate. Chemical Society Reviews
61. United States Environmental Protection Agency (2016) Integrated science assessment for carbon monoxide
62. Beelen R, et al. (2014) Effects of long-term exposure to air pollution on natural-cause mortality: an analysis of 22 European cohorts within the multicentre ESCAPE project. The Lancet
63. Lehmann S (2004) Topography and its effects on air pollution in urban areas. Urban Climate
64. Brauer M, et al. (2012) Exposure assessment for estimation of the global burden of disease attributable to outdoor air pollution. Environ Sci Technol
65. Harrison RM, et al. (1997) Measurements of the physical properties of particles in the urban atmosphere. Atmos Environ
66. Vingarzan R (2004) A review of surface ozone background levels and trends. Atmospheric Environment
67. Gupta P, Christopher SA (2009) Particulate matter air quality assessment using integrated surface, satellite, and meteorological products: Multiple regression approach. J Geophys Res
68. Kim S, et al. (2006) Air pollution and health impacts in China: A review. Environ Int
69. Jacob DJ, Winner DA (2009) Effect of climate change on air quality. Atmos Environ
70. Intergovernmental Panel on Climate Change (IPCC) (2013) Climate change 2013: The physical science basis
71. Zhang Q, et al. (2015). Reducing PM2.5 in China: Results from a source-oriented chemical transport model. Environ Sci Technol
72. Fenger J (1999) Urban air quality. Atmospheric Environment
73. Kumar P, et al. (2015) Winter air pollution from biomass burning in northern India: An overview. Environ Res
74. Smith E, Johnson T (2019) Industrial emission control technologies: A review. Environ Eng J 84(2):117–129
75. Brown A, Miller D, Green P (2020) Process optimization in industrial emission control. Ind Eng J 49(3):201–213
76. Environmental Agency (2021) Regulatory enforcement and compliance in air quality management. Retrieved from www.environmental-agency.gov
77. Green M, Miller L (2018) The role of electric vehicles in reducing urban air pollution. Transp Res Part D: Transp Environ 64:24–34
78. World Health Organization (WHO) (2019) Fuel quality and emission control. Retrieved from www.who.int
79. Anderson J, Lee C (2020) Traffic management strategies for reducing urban air pollution. J Environ Planning Manage 63(5):723–737
80. Johnson T, Smith R, Williams D (2017) Energy-efficient appliances and air quality improvement. Energy Policy 108:457–467
81. Environmental Protection Agency (EPA) (2020) Regulation of household products to control indoor air pollution. Retrieved from www.epa.gov
82. Chen H, Zhao J (2020) Advancements in air quality monitoring technologies. Environ Sci Technol 54(10):6073–6082
83. Li X, Wang Y, Zhou Z (2019) Data analysis and atmospheric modelling for air quality management. Atmos Environ 213:413–423
84. Gupta A, Sharma M, Singh N (2018) Development of emission inventories for urban air quality management. Atmos Environ 186:21–32
85. Environmental Protection Agency (EPA) (2021) Clean air act and national ambient air quality standards. Retrieved from www.epa.gov

86. Smith R (2021) Penalties and Enforcement in Environmental Regulations. Environ Policy Rev 27(1):32–45
87. Jones P, Baker S (2019) Community monitoring programs and citizen science in air quality assessment. Environ Monit Assess 191(4):223
88. Dunn K, Henneberry J (2020) Green Infrastructure: A planning and design guide. CABI
89. Carmona M, Magalhães CD (2018) Public places, urban spaces: The dimensions of urban design. Routledge
90. Banister D, Hickman R (2019) Transport futures: Exploring the future of transport and the environment. Oxford University Press
91. Pucher J, Buehler R (2017) City Cycling. MIT Press
92. Gehl J (2018) Cities for People. Island Press
93. Beatley T (2017) Handbook of Biophilic City Planning & Design. Island Press
94. Tan PY (2016) Sustainable Urban Forest Management. Springer
95. Kumar P, Omidvarborna H, Sharma C (2019) The rise of low-cost sensing for managing air pollution in cities. Environ Int 126:580–589
96. Castell N, Kobernus M, Liu H-Y, Schneider P, Lahoz W, Berre AJ, Noll J (2017) Mobile technologies and services for environmental monitoring: The Citi-Sense-MOB approach. Urban Climate 20:203–212
97. Duncan BN, Yoshida Y, Damon M, Douglass AR, Witte JC, Koumoutsaris S, Frith SM (2016) Space-based observations of air quality: Progress and promise. J Air Waste Manag Assoc 66(9):885–907
98. Cheng YH, Tsai CJ, Lai CC, Wang YN (2017) Performance of a photocatalytic filter reactor coated with TiO2 for abating VOC emissions. Environ Technol 38(15):1843–1850
99. Guerra R, Caetano MAL, González A (2021) Vertical gardens as an urban farming alternative for climate change mitigation. Agric Syst 191:103141
100. Hänninen OO, Knol AB, Jantunen M, Lim TA, Conrad A, Rappolder M, Bartnett AG (2014) Environmental burden of disease in Europe: assessing nine risk factors in six countries. Environ Health Perspect 122(5):439–446
101. Ramaswami A, Russell AG, Culligan PJ, Sharma KR, Kumar E (2016) Meta-principles for developing smart, sustainable, and healthy cities. Science 352(6288):940–943
102. Creutzig F, Jochem P, Edelenbosch OY, Mattauch L, Vuuren DPV, McCollum D, Minx J (2015) Transport: A roadblock to climate change mitigation? Science 350(6263):911–912
103. International Energy Agency (IEA) (2020) Global EV Outlook 2020: Entering the decade of electric drive? Retrieved from www.iea.org
104. O'Neill MS, Carter R, Kish JK, Gronlund CJ, White-Newsome JL, Manarolla X, Schwartz J (2020) Preventing heat-related morbidity and mortality: New approaches in a changing climate. Maturitas 136:10–17

Environmental Degradation and Impacts on Agricultural Production: A Challenge to Urban Sustainability

Harshit Mishra

Abstract Sustainable urban growth presents considerable issues due to its effects on agricultural productivity and environmental degradation. Agricultural land is converted as a result of urban sprawl, which also fragments landscapes and lowers the ecosystem services that these places offer. Because of pollution from urban runoff, cities and agriculture both compete for the same resources-water-both in terms of quantity and quality. Another issue is air quality, since urban emissions have an effect on agricultural output. These problems are made worse by climate change, which increases agriculture's susceptibility to extreme weather events. Production in agriculture is directly impacted by environmental degradation. Unsustainable land-use practices deplete soil fertility by contaminating and eroding it. Pollution and lack of water put additional strain on agricultural production due to salinization, competition for resources and contaminated water. Agriculture depends on natural processes like pollination, which are disrupted by biodiversity loss. Climate change increases the frequency and intensity of droughts and floods, making agricultural systems more susceptible to shocks. It also brings about unpredictable precipitation patterns and rising temperatures. Effective governance and policies are required to address these issues. It is essential to use cutting-edge urban agriculture methods, technical solutions to slow down environmental deterioration and technology integration to create sustainable urban habitats. To encourage knowledge, behaviour change and community involvement in sustainable practices, education and community engagement are essential. Creating market incentives for a more sustainable future requires an understanding of the financial costs of unsustainable activities and the advantages of sustainable alternatives.

Keywords Agricultural production · Climate change · Environmental degradation · Sustainability · Urban sprawl

H. Mishra (✉)
Department of Agricultural Economics, College of Agriculture, Acharya Narendra Deva University of Agriculture and Technology, Kumarganj, Ayodhya (U.P.) 224 229, India
e-mail: wehars@gmail.com

J. A. Parray et al. (eds.), *Sustainable Urban Environment and Waste Management*, Urban Sustainability, https://doi.org/10.1007/978-981-96-1140-9_3

1 Introduction

Agricultural production, the cornerstone of human civilization, faces a multitude of challenges in the twenty-first century. Feeding a growing global population projected to reach 9.7 billion by 2050, while navigating the treacherous terrain of environmental degradation, presents an unprecedented test for our agricultural systems [1, 2]. This challenge is further amplified by the relentless march of urbanization. Cities are currently home to over half the world's population, with projections suggesting this number will balloon to 68% by 2050. This rapid urban sprawl not only gobbles up fertile agricultural land, but also disrupts ecosystems vital for healthy food production [3]. A 2020 study by the United Nations Food and Agriculture Organization (FAO) estimates that by 2050, we'll need to increase global food production by 60% just to keep pace with population growth, a daunting task considering the ongoing loss of arable land [4].

Environmental degradation adds another layer of complexity. Climate change, driven by greenhouse gas emissions largely from human activities, is wreaking havoc on agricultural productivity. Rising temperatures, erratic precipitation patterns and increased frequency of extreme weather events like floods and droughts are disrupting agricultural cycles and reducing crop yields [5, 6]. A 2019 report by the Intergovernmental Panel on Climate Change (IPCC) warns that these trends, if left unchecked, could lead to yield reductions of major crops like maize and wheat by as much as 7% per decade. The consequences of these interconnected challenges are far-reaching. Food security, the ability of all people to access sufficient, safe and nutritious food, is under immense pressure. The FAO estimates that in 2020, around 750 million people were undernourished, a stark reminder of the precarious state of global food systems. This problem is likely to worsen as agricultural production struggles to keep up with demand, potentially leading to social unrest and political instability [7–9].

The spectre of malnutrition also looms large. A 2020 study published in the Lancet medical journal found that deficiencies in essential vitamins and minerals (micronutrient deficiencies) affect an estimated two billion people globally, with the most severe cases concentrated in low- and middle-income countries. This not only has a debilitating impact on individual health, but also hinders cognitive development and economic productivity [10]. Furthermore, the environmental cost of conventional agricultural practices is becoming increasingly unsustainable. Intensive farming techniques, characterized by heavy reliance on chemical fertilizers and pesticides, have led to soil erosion, water pollution and biodiversity loss. A 2017 report by the World Wildlife Fund (WWF) found that agriculture is the leading cause of deforestation globally and a major contributor to greenhouse gas emissions [11, 12].

The urgency of the situation demands a paradigm shift in the way we produce food. Sustainable agriculture practices, which focus on protecting the environment while ensuring food security, offer a glimmer of hope. These practices include crop rotation, integrated pest management and the use of cover crops to improve soil health and fertility. Additionally, advancements in agricultural technology, such as precision agriculture and vertical farming, hold promise for optimizing resource use and increasing yields [13]. However, the transition towards a more sustainable food system is not without its challenges. Investment in research and development is crucial to improve existing techniques and develop new ones. Additionally, there's a need to address social and economic barriers that hinder the adoption of sustainable practices by small-scale farmers, who are often the most vulnerable to environmental degradation and climate change [14].

Finally, promoting consumer awareness and encouraging dietary shifts towards more plant-based diets can significantly reduce the environmental footprint of our food systems. A 2018 study published in Science found that widespread adoption of plant-based diets could reduce food-related greenhouse gas emissions by up to 50%. The challenges facing agricultural production in the twenty-first century are daunting [15, 16]. However, by embracing sustainable practices, investing in innovation and fostering global cooperation, we can build a more resilient and equitable food system that nourishes a growing population while protecting the environment for generations to come. Keeping in view the above facts, the present study is specifically undertaken and carried out with the following objectives:

- To analyse the impact of urban sprawl on agricultural land and productivity.
- To assess the potential of urban waste management practices to support sustainable agriculture.
- To evaluate the vulnerability of agricultural systems to climate change and extreme weather events.

2 Challenges to Urban Sustainability

Urban areas face multifaceted challenges that pose significant obstacles to achieving sustainability objectives. This section explores four key challenges confronting urban sustainability: land use and urban sprawl, water resource management, air quality and climate change and waste management and resource recovery. Each challenge is examined in detail to elucidate its implications for urban environments and the broader sustainability agenda [17].

2.1 Land Use and Urban Sprawl

Urban sprawl, characterized by the unplanned, unrestricted expansion of urban areas into surrounding rural landscapes, presents a significant impediment to sustainable urban development. The following subsections elucidate the major challenges associated with land use dynamics and urban sprawl:

2.1.1 Conversion of Agricultural Land for Urban Development

The conversion of agricultural land for urban development represents a significant threat to both agricultural production and overall environmental sustainability. As urban populations burgeon, there is an escalating demand for residential, commercial and industrial spaces, leading to the encroachment of urban sprawl into rural areas. This encroachment often results in the irreversible loss of fertile agricultural land, diminishing the capacity for local food production and exacerbating dependency on distant sources [18, 19].

The ramifications of agricultural land conversion extend beyond mere loss of arable land. It disrupts local ecosystems, alters hydrological regimes and impairs biodiversity by fragmenting habitats. Additionally, the conversion of agricultural land contributes to soil degradation, exacerbating erosion and reducing soil fertility. Such alterations in land use patterns can disrupt the delicate balance of ecosystems, leading to cascading ecological consequences.

Policy interventions aimed at curbing agricultural land conversion necessitate a comprehensive approach encompassing land-use planning, zoning regulations and incentives for brownfield redevelopment. Furthermore, the promotion of compact urban development models and the implementation of greenbelt policies can help mitigate the relentless encroachment of urban sprawl into agricultural hinterlands [20].

2.1.2 Fragmentation of Agricultural Landscapes

The fragmentation of agricultural landscapes, resulting from urban sprawl and infrastructural development, poses formidable challenges to agricultural sustainability and ecosystem resilience. Fragmentation disrupts the contiguous expanse of agricultural land, leading to the subdivision of once-cohesive landscapes into smaller, isolated patches. This fragmentation diminishes the viability of agricultural operations, impeding economies of scale and exacerbating logistical challenges [21].

The adverse effects of landscape fragmentation extend beyond the spatial domain, permeating ecological processes and biodiversity dynamics. Fragmented landscapes impede the movement of flora and fauna, fragmenting habitats and hindering gene flow among populations [22]. Consequently, this impediment to ecological connectivity increases the vulnerability of species to extinction and compromises the resilience of ecosystems in the face of environmental stressors.

Addressing the fragmentation of agricultural landscapes necessitates a multifaceted approach that integrates land-use planning, habitat restoration and ecological corridors. By prioritizing the preservation of contiguous agricultural expanses and implementing measures to enhance landscape connectivity, policymakers can foster agricultural sustainability while concurrently safeguarding biodiversity and ecosystem resilience [23, 24].

2.1.3 Loss of Ecosystem Services Provided by Agricultural Lands

Agricultural lands are not merely sites of food production but also serve as critical providers of ecosystem services essential for environmental sustainability and human well-being. The conversion and degradation of agricultural lands precipitate the loss of these invaluable ecosystem services, jeopardizing the resilience of both rural and urban landscapes.

One of the primary ecosystem services provided by agricultural lands is soil fertility maintenance and nutrient cycling. However, the conversion of agricultural land for urban development disrupts these processes, leading to soil degradation, erosion and nutrient depletion. Consequently, this degradation diminishes the capacity of soils to support agricultural productivity, necessitating intensive inputs of fertilizers and agrochemicals [25–27].

Moreover, agricultural landscapes play a pivotal role in regulating hydrological cycles and mitigating the impacts of floods and droughts. The loss of vegetative cover resulting from land conversion diminishes the capacity of agricultural lands to intercept rainfall, exacerbating soil erosion and flooding. Furthermore, the degradation of wetlands and riparian buffers diminishes the capacity of agricultural landscapes to attenuate water pollution and maintain water quality.

The preservation of agricultural lands is imperative for ensuring both urban sustainability and environmental resilience. Addressing the challenges of land use conversion, landscape fragmentation and the loss of ecosystem services necessitates a concerted effort encompassing policy interventions, land-use planning strategies and stakeholder engagement [28]. By prioritizing the preservation and sustainable management of agricultural landscapes, policymakers can foster a harmonious coexistence between urban development and environmental conservation, ensuring the long-term viability of agricultural production systems [29].

2.2 *Water Resource Management*

Water resource management stands as a critical facet in the pursuit of urban sustainability, particularly as urbanization intensifies and challenges associated with water availability, quality and allocation become more pronounced. Within this domain, several key challenges emerge, each necessitating meticulous attention and strategic intervention.

2.2.1 Increased Water Demand in Cities

The burgeoning population within urban centres exerts substantial pressure on water resources, driving a significant surge in water demand. This escalating demand stems from various factors including population growth, industrial expansion and heightened standards of living. As urban populations swell, so too does the need for water for domestic, industrial and commercial purposes. Consequently, municipal water utilities often struggle to meet this escalating demand, leading to issues such as water shortages, supply inadequacies and increased reliance on alternative sources such as groundwater [30, 31].

Addressing this challenge mandates a multifaceted approach encompassing both demand-side management strategies and supply augmentation measures. Demand-side management initiatives may include the promotion of water-efficient technologies, implementation of conservation measures and the establishment of public awareness campaigns to foster responsible water usage practices. Additionally, supply augmentation strategies could entail the development of alternative water sources such as reclaimed wastewater, rainwater harvesting systems and desalination plants. Collaborative efforts between governmental bodies, urban planners, water authorities and the private sector are imperative to navigate the complexities associated with managing increased water demand in cities effectively [32].

2.2.2 Pollution of Water Resources From Urban Runoff

Urban runoff, characterized by the discharge of stormwater and pollutants from urbanized landscapes into water bodies, poses a significant threat to water resource quality and ecosystem integrity. The impervious surfaces prevalent in urban environments exacerbate the issue by accelerating the rate of surface water runoff, thereby carrying a plethora of pollutants—including heavy metals, hydrocarbons, nutrients and pathogens—into nearby waterways. Consequently, urban runoff engenders a myriad of adverse environmental ramifications such as water contamination, habitat degradation and the impairment of aquatic ecosystems.

Combatting the pollution stemming from urban runoff necessitates the implementation of comprehensive stormwater management strategies tailored to mitigate pollutant loads and minimize environmental degradation [33]. Green infrastructure initiatives, encompassing practices such as permeable pavement, green roofs and constructed wetlands, serve as effective mechanisms for attenuating the impacts of urban runoff by facilitating natural filtration, groundwater recharge and flood mitigation. Furthermore, regulatory frameworks mandating the implementation of stormwater best management practices (BMPs) and the enforcement of pollutant discharge limits play a pivotal role in safeguarding water quality and promoting sustainable urban development [34, 35].

2.2.3 Competition for Water Between Urban and Agricultural Uses

The confluence of urbanization and agricultural intensification engenders a scenario wherein competition for water resources between urban and agricultural sectors becomes increasingly pronounced. This competition is predicated on the finite nature of water resources juxtaposed with the escalating demands emanating from both urban and agricultural domains. Urban areas, characterized by burgeoning populations and burgeoning industrial activities, vie for water to sustain domestic needs, industrial processes and commercial endeavours. Conversely, agriculture, as a fundamental pillar of food security and economic stability, necessitates substantial water allocations to irrigate crops, sustain livestock and bolster agricultural productivity [36].

This intersectoral competition for water engenders a complex dynamic fraught with socio-economic, environmental and political ramifications. Disparities in water access and allocation exacerbate existing inequities, disproportionately impacting marginalized communities and exacerbating socio-economic disparities. Moreover, heightened water withdrawals for urban and agricultural purposes exert pressure on freshwater ecosystems, leading to habitat degradation, biodiversity loss and ecosystem destabilization [37]. In navigating this intricate web of challenges, equitable water governance frameworks, incorporating principles of integrated water resource management (IWRM) and stakeholder engagement, are imperative to foster dialogue, promote cooperation and ensure the sustainable utilization of water resources for both urban and agricultural needs.

2.3 Air Quality and Climate Change

Urban areas are significant contributors to air pollution and climate change, posing significant challenges to agricultural productivity. This section explores the specific impacts of urban emissions and climate change events on agriculture, along with the role of urban planning in mitigating these impacts.

2.3.1 Urban Emissions Impacting Agricultural Productivity

Urban emissions encompass a spectrum of pollutants emitted from various anthropogenic activities within urban centres (as shown in Fig. 1). These emissions include, but are not limited to, particulate matter (PM), nitrogen oxides (NOx), sulphur dioxide (SO_2), volatile organic compounds (VOCs) and carbon monoxide (CO). Their detrimental effects on agricultural productivity are manifold [38].

Firstly, these emissions contribute to the formation of ground-level ozone (O_3) and smog, which can directly harm crops by inhibiting photosynthesis, reducing crop yields and damaging plant tissues. Ozone, in particular, interferes with the plant's ability to absorb carbon dioxide and thus affects its growth and development. Secondly, pollutants such as NOx and VOCs can undergo chemical reactions in the atmosphere to form secondary pollutants like peroxyacetyl nitrate (PAN) and ozone, which are known to cause oxidative stress in plants, leading to reduced growth, flowering and seed production.

Fig. 1 City smog reduces sunlight, harming crops (Urban emissions can stunt plant growth and lower agricultural yields)

Moreover, particulate matter, especially PM2.5 (particles with a diameter of 2.5 mm or less), can settle on plant surfaces, blocking stomata, reducing photosynthesis and impairing nutrient uptake. This not only diminishes crop yields but also affects the quality and nutritional value of agricultural produce [39].

2.3.2 Vulnerability of Agriculture to Climate Change Events

Climate change poses significant challenges to agricultural production, exacerbating existing vulnerabilities and introducing new risks. Urbanization often amplifies these challenges due to the concentration of infrastructure, populations and economic activities in urban areas.

Extreme weather events, such as heatwaves, droughts, floods and storms, are becoming more frequent and intense under climate change scenarios. These events can directly damage crops, disrupt planting and harvesting schedules and lead to soil erosion and nutrient loss.

Furthermore, shifts in temperature and precipitation patterns alter the suitability of certain regions for specific crops, forcing farmers to adapt by changing crop varieties or farming practices. Urban sprawl encroaches upon arable land, limiting agricultural expansion and exacerbating competition for resources such as water and land [40].

2.3.3 The Role of Urban Planning in Mitigating Climate Impact on Agriculture

Urban planning plays a pivotal role in mitigating the adverse effects of climate change on agriculture by promoting sustainable land use practices, enhancing resilience and reducing greenhouse gas emissions.

Integrated land-use planning can help preserve agricultural land by designating protected areas or implementing zoning regulations to prevent the conversion of agricultural land for urban development. Additionally, promoting urban agriculture and green spaces within cities can contribute to food security, mitigate urban heat island effects and improve air quality. Incorporating climate-resilient design principles into urban infrastructure can minimize the vulnerability of both urban and peri-urban agricultural systems to extreme weather events. This may include implementing green infrastructure solutions such as permeable pavements, rain gardens and green roofs to mitigate flooding and enhance water retention.

Furthermore, incentivizing low-carbon transportation modes and promoting energy-efficient buildings can reduce urban emissions, mitigating the contribution of cities to climate change and its indirect impacts on agriculture [41]. Addressing the challenges posed by urban emissions and climate change requires a multifaceted approach that integrates urban planning, sustainable development strategies and climate resilience measures to safeguard agricultural productivity and ensure the long-term sustainability of food systems in urbanizing regions.

2.4 Waste Management and Resource Recovery

Effective waste management and resource recovery are critical components. Urban centres generate substantial amounts of waste, posing significant challenges to environmental sustainability and public health. This section explores the complexities of managing urban waste, explores the potential for utilizing urban waste in sustainable agricultural practices and discusses circular economy approaches that integrate waste management with urban-agricultural systems.

2.4.1 Challenges of Managing Urban Waste

Urban waste management presents a multifaceted challenge in the context of urban sustainability. The burgeoning urban population, coupled with rapid urbanization, has led to an exponential increase in the generation of waste, encompassing various forms such as municipal solid waste, electronic waste and industrial waste. The management of this diverse array of waste streams poses significant challenges due to several interconnected factors. Firstly, the sheer volume of urban waste generated overwhelms existing waste management infrastructures and capacities. In many urban areas, inadequate waste collection, transportation and disposal facilities result in the accumulation of waste in streets, vacant lots and water bodies, contributing to environmental degradation and public health hazards. Secondly, the composition of urban waste has become increasingly complex, containing a mix of organic, inorganic and hazardous materials. This heterogeneity complicates waste management processes, as different types of waste require specialized handling and treatment methods. Moreover, the presence of hazardous substances poses risks to both human health and the environment if not managed properly [42, 43].

Additionally, the lack of awareness and apathy towards waste management practices among urban residents exacerbates the challenge. Inadequate segregation of waste at the source leads to contamination and hampers recycling and resource recovery efforts. Furthermore, cultural attitudes towards waste disposal, including littering and informal dumping, perpetuate the problem and hinder sustainable waste management initiatives. Furthermore, financial constraints and governance issues often impede the implementation of comprehensive waste management strategies. Limited funding for infrastructure development and maintenance, coupled with bureaucratic inefficiencies and political inertia, hamper efforts to improve waste management systems and enact appropriate policies and regulations.

Addressing these challenges requires a holistic approach that integrates technological innovation, policy reforms, public engagement and institutional capacity building. By enhancing waste collection and recycling infrastructure, promoting public awareness and participation, enforcing regulations and fostering collaboration between stakeholders, urban areas can strive towards more sustainable waste management practices [44].

2.4.2 Potential for Using Urban Waste in Sustainable Agriculture

The utilization of urban waste in sustainable agriculture presents a promising opportunity to address both environmental and agricultural challenges. Urban waste, particularly organic waste derived from households, restaurants and food processing industries, contains valuable nutrients and organic matter that can be recycled and utilized as soil amendments and fertilizers in agricultural production. One of the major benefits of using urban waste in agriculture is its potential to enhance soil fertility and productivity. Organic waste, such as kitchen scraps, yard trimmings and sewage sludge, contains essential nutrients such as nitrogen, phosphorus and potassium, which are vital for plant growth. By composting organic waste or converting it into biofertilizers, urban areas can reduce the reliance on synthetic fertilizers derived from non-renewable resources and mitigate the environmental impacts associated with their production and use.

Moreover, integrating urban waste into agricultural systems promotes circularity and resource efficiency by closing the nutrient loop. Rather than disposing of organic waste in landfills or incinerators, recycling it back into the soil replenishes nutrient stocks and improves soil structure, thereby supporting sustainable crop production and mitigating soil degradation. Furthermore, the use of urban waste in agriculture can contribute to the reduction of greenhouse gas emissions and mitigate climate change [45]. By diverting organic waste from landfills, where it decomposes anaerobically and produces methane, a potent greenhouse gas, to composting or anaerobic digestion facilities, emissions can be reduced. Additionally, incorporating organic matter into soil helps sequester carbon, thereby offsetting emissions and enhancing soil carbon stocks. However, the potential for using urban waste in sustainable agriculture is contingent upon overcoming several challenges. These include concerns regarding the quality and safety of recycled waste products, such as heavy metal contamination and pathogens, which may pose risks to human health and the environment if not properly managed. Moreover, logistical issues related to waste collection, transportation and processing need to be addressed to ensure the efficient and cost-effective utilization of urban waste in agriculture.

Policy and regulatory frameworks play a crucial role in incentivizing the use of urban waste in agriculture while safeguarding human health and environmental integrity. Establishing quality standards, monitoring systems and certification schemes can help ensure the safety and efficacy of recycled waste products. Furthermore, providing financial incentives and technical support to farmers and waste management enterprises can encourage the adoption of sustainable waste-to-agriculture practices. Leveraging urban waste in sustainable agriculture offers a viable pathway towards enhancing urban sustainability, improving resource efficiency and promoting circular economy principles. By harnessing the nutrient and organic matter content of urban waste, cities can contribute to the resilience and sustainability of agricultural systems while simultaneously addressing urban waste management challenges [46, 47].

2.4.3 Circular Economy Approaches for Waste Management in Urban-Agricultural Systems

The concept of a circular economy offers a holistic and integrated approach to waste management in urban-agricultural systems, emphasizing the minimization of waste generation, resource conservation and the promotion of resource efficiency and value retention throughout the product life cycle. Central to circular economy approaches is the principle of closing the loop by designing out waste and pollution, keeping products and materials in use for as long as possible and regenerating natural systems. In the context of urban-agricultural systems, circular economy strategies seek to create synergies and symbiotic relationships between urban and agricultural activities, thereby optimizing resource utilization and minimizing environmental impacts. One key circular economy approach for waste management in urban-agricultural systems is the concept of urban metabolism, which views cities as complex ecosystems that consume resources, generate waste and interact with their surrounding environments. By analysing the flow of materials and energy within urban systems, policymakers and planners can identify opportunities for resource recovery, recycling and waste valorisation [48].

Moreover, integrating agricultural activities into urban landscapes, such as urban farming, rooftop gardening and peri-urban agriculture, can facilitate the recycling of organic waste and the production of locally grown food, thereby reducing the ecological footprint of food production and distribution. Furthermore, the adoption of circular business models, such as product-as-a-service, remanufacturing and sharing platforms, can promote resource efficiency and waste reduction in urban-agricultural systems. By shifting from a linear "take-make-dispose" model to a circular "reduce-reuse-recycle" paradigm, businesses can minimize waste generation, extend product lifespans and create value from waste streams. Additionally, technological innovation plays a crucial role in enabling circular economy approaches for waste management in urban-agricultural systems [49]. Advanced waste treatment technologies, such as anaerobic digestion, composting and bioconversion, can facilitate the conversion of organic waste into valuable products, such as biogas, compost and biofertilizers. Likewise, digital technologies, such as blockchain and internet-of-things (IoT) sensors, can enhance traceability, transparency and efficiency in waste management and resource recovery processes.

However, realizing the full potential of circular economy approaches for waste management in urban-agricultural systems requires overcoming various barriers and challenges. These include regulatory barriers, market distortions, technological limitations and institutional constraints, which may hinder the adoption and implementation of circular economy initiatives. Therefore, concerted efforts are needed from policymakers, businesses, civil society and academia to create an enabling environment for circular economy practices in urban-agricultural systems. This entails fostering collaboration and knowledge exchange, providing financial incentives and technical assistance, raising awareness and building capacity and mainstreaming circular economy principles into urban planning and development strategies. Circular economy approaches offer a promising framework for addressing the challenges of waste management in urban-agricultural systems while promoting sustainability, resilience and resource efficiency [50, 51].

3 Environmental Degradation and Its Impact on Agriculture

Urbanization and the associated environmental degradation pose significant challenges to agricultural production. This section delineates the multifaceted ways in which environmental degradation impacts agriculture, focusing on soil degradation and loss, water scarcity and pollution, biodiversity loss and ecosystem services, as well as climate change and extreme events.

3.1 Soil Degradation and Loss

Soil, the fundamental resource for agriculture, is under severe threat due to urbanization-induced environmental degradation. This subsection dissects the various facets of soil degradation and loss, encompassing erosion, contamination and loss of fertility and productivity.

3.1.1 Erosion Due to Unsustainable Land-Use Practices

Erosion stands as a critical consequence of unsustainable land-use practices within urban environments, posing a formidable threat to agricultural productivity. Unsustainable land-use practices such as deforestation, improper land management and excessive construction without adequate erosion control measures exacerbate soil erosion. In urban areas, the proliferation of impervious surfaces, such as roads and buildings, exacerbates surface runoff, leading to increased soil erosion. The removal of vegetative cover, typically through deforestation or urbanization, exposes soil to the erosive forces of wind and water. Without the protective cover of vegetation, soil becomes susceptible to erosion, as rainwater washes away the topsoil and exacerbates soil degradation (as shown in Fig. 2). Moreover, the construction of infrastructure, such as roads and buildings, often involves the disturbance of soil, further accelerating erosion rates [52].

Soil erosion not only results in the physical loss of soil but also leads to the degradation of soil structure and fertility. As fertile topsoil is eroded away, the remaining soil becomes less capable of supporting agricultural productivity, thus impeding sustainable agricultural practices within urban areas. Furthermore, the sediment

Fig. 2 Parched earth with sparse plants reveals erosion (This land suffers from unsustainable practices, losing valuable topsoil)

runoff from erosion can clog waterways and degrade aquatic ecosystems, exacerbating the environmental impact beyond agricultural lands. Addressing erosion due to unsustainable land-use practices necessitates the implementation of erosion control measures, including the adoption of sustainable land management practices, reforestation efforts and the construction of erosion control structures such as terraces and retaining walls. Additionally, zoning regulations and urban planning strategies should prioritize the preservation of vegetative cover and the integration of green spaces within urban landscapes to mitigate erosion risks and uphold agricultural sustainability.

3.1.2 Soil Contamination from Urban Pollutants

Soil contamination stemming from urban pollutants poses a significant threat to agricultural production and environmental integrity, highlighting the intricate relationship between urbanization and agricultural sustainability. Urban areas are characterized by a myriad of anthropogenic activities, including industrial operations, transportation networks and residential developments, which collectively contribute to the release of various pollutants into the environment [53]. Urban pollutants, such as heavy metals, petroleum hydrocarbons, pesticides and industrial chemicals, can infiltrate soil through direct deposition or atmospheric deposition, contaminating agricultural lands and compromising soil quality. Industrial activities, in particular, are primary sources of soil contamination in urban environments, as improper waste disposal practices and accidental spills can result in the release of hazardous substances into the soil.

The presence of contaminants in soil can have detrimental effects on agricultural productivity, crop quality and human health. Heavy metals, such as lead, cadmium and mercury, can accumulate in soil and subsequently be taken up by plants, leading to food chain contamination and posing risks to human consumers. Moreover, soil contamination can impair soil fertility, disrupt microbial activity and inhibit plant growth, thereby diminishing agricultural yields and compromising food security. Mitigating soil contamination from urban pollutants requires a multifaceted approach encompassing pollution prevention, remediation efforts and regulatory measures. Implementing pollution control technologies in industrial facilities, enforcing strict environmental regulations and promoting sustainable waste management practices are essential steps in reducing the release of pollutants into the environment. Additionally, remediation techniques such as soil excavation, phytoremediation and soil amendments can help mitigate existing soil contamination and restore soil health for agricultural purposes [54].

3.1.3 Loss of Soil Fertility and Productivity

The loss of soil fertility and productivity stands as a pressing challenge confronting agricultural sustainability in urban areas, driven by a multitude of factors including soil degradation, contamination and inadequate land management practices. Soil fertility, defined as the soil's capacity to support plant growth and sustain agricultural productivity, is a fundamental determinant of agricultural success and food security. Urbanization exerts significant pressure on soil fertility and productivity through various mechanisms, including soil erosion, contamination, compaction and loss of organic matter. Soil erosion, as discussed previously, results in the physical removal of fertile topsoil, which is rich in organic matter, nutrients and beneficial microorganisms essential for plant growth. Contaminants introduced into the soil can further degrade soil fertility by altering its chemical composition and inhibiting nutrient uptake by plants [55, 56].

Compaction, another consequence of urbanization, occurs due to heavy machinery use, pedestrian traffic and construction activities, leading to the compression of soil particles and the reduction of pore space essential for air and water movement. Compacted soils exhibit poor aeration, drainage and root penetration, impairing plant growth and productivity. Furthermore, the loss of organic matter, resulting from urban development and intensive land-use practices, diminishes soil fertility by depriving the soil of essential nutrients and organic carbon necessary for sustaining microbial activity and nutrient cycling.

Addressing the loss of soil fertility and productivity requires concerted efforts aimed at soil conservation, restoration and sustainable land management practices. Implementing soil conservation practices such as cover cropping, crop rotation and reduced tillage can help mitigate soil erosion, improve soil structure and enhance soil fertility. Additionally, integrating organic amendments, such as compost and manure, into soil can replenish organic matter and nutrient content, fostering soil health and agricultural productivity [57]. Promoting agroecological approaches that prioritize biodiversity, soil conservation and ecosystem resilience is essential for ensuring the long-term sustainability of agricultural systems in urban environments. These strategies should be complemented by education, outreach and policy initiatives aimed at raising awareness about the importance of soil conservation and fostering stewardship of urban agricultural lands. By addressing the underlying drivers of soil degradation and promoting sustainable land management practices, urban areas can uphold agricultural sustainability, mitigate environmental degradation and secure food production for future generations [58].

3.2 Water Scarcity and Pollution

Water scarcity and pollution represent pressing concerns for agricultural sustainability, affecting water availability, quality and ecosystem health.

3.2.1 Competition for Water Resources Between Urban and Agricultural Uses

Water, as a finite resource, is increasingly under pressure due to the growing demands of both urban and agricultural sectors. In many regions, urban expansion encroaches upon agricultural land, leading to heightened competition for limited water resources. Urban areas, driven by population growth and industrialization, often prioritize water for domestic, industrial and recreational purposes. This allocation can strain available water supplies, leaving less for agricultural needs. Moreover, the infrastructure developments accompanying urban growth, such as reservoirs and diversion channels, can further reduce water availability for agriculture. These structures redirect water away from rural areas towards urban centres, exacerbating the competition. Additionally, policies and regulations may favour urban water users, leading to unequal access and distribution.

The consequences of this competition are significant for agricultural production. Farmers may face reduced access to water for irrigation, impacting crop yields and quality. Furthermore, increased competition can drive up water prices, rendering irrigation practices economically unfeasible for small-scale farmers. Ultimately, this competition threatens the sustainability of agricultural systems and food security, especially in regions where agriculture remains a primary livelihood [59].

3.2.2 Salinization and Waterlogging Due to Irrigation Practices

Irrigation, while essential for agricultural productivity, can inadvertently contribute to soil salinization and waterlogging, particularly in arid and semi-arid regions. Intensive irrigation practices, coupled with poor drainage systems, can lead to the accumulation of salts in the soil profile. As water evaporates from the surface, salts become increasingly concentrated, hindering plant growth and diminishing soil fertility (as shown in Fig. 3). Similarly, waterlogging occurs when excessive irrigation or inadequate drainage saturates the soil, displacing oxygen and impeding root respiration. This anaerobic condition can lead to root suffocation and nutrient imbalances, further compromising crop health and productivity [60].

The consequences of salinization and waterlogging are profound for agricultural production. Affected soils become less suitable for cultivation, leading to reduced yields and crop failures. Moreover, the economic costs associated with soil remediation and drainage infrastructure can be prohibitive for farmers, particularly in resource-constrained settings. Thus, addressing these challenges requires integrated water management strategies that balance irrigation efficiency with soil conservation measures.

Fig. 3 Salt-crusted field shows irrigation's impact on agriculture field (The white substance on the soil is salt left behind by evaporated irrigation water. This can harm crops and reduce soil fertility)

3.2.3 Impact of Water Pollution on Agricultural Production

Water pollution poses a significant threat to agricultural production, jeopardizing both crop health and food safety. Urban activities, including industrial discharge, untreated sewage and runoff from urban areas, introduce a plethora of contaminants into water bodies (as shown in Fig. 4). These pollutants include heavy metals, pathogens, pesticides and synthetic chemicals, which can accumulate in soil and water resources [61]. When contaminated water is used for irrigation, crops can absorb and accumulate harmful substances, posing risks to human health upon consumption. Additionally, polluted water sources may directly harm crops by stunting growth, reducing yields and inducing physiological disorders. Furthermore, water pollution can disrupt aquatic ecosystems, affecting biodiversity and ecosystem services critical for agriculture, such as pollination and pest control.

Addressing the impact of water pollution on agricultural production necessitates comprehensive strategies encompassing pollution prevention, water treatment and sustainable agricultural practices. Efforts to mitigate pollution at its source, such as stricter regulations and wastewater treatment infrastructure, are essential to safeguarding water quality. Moreover, promoting agroecological approaches that reduce reliance on synthetic inputs and enhance soil health can mitigate the adverse effects of pollution on crop productivity and food safety [62].

3.3 Biodiversity Loss and Ecosystem Services

Biodiversity loss, exacerbated by various anthropogenic activities, poses a significant threat to agricultural sustainability and ecosystem resilience. This section explores

Fig. 4 Industrial runoff pollutes river, harming agriculture (Polluted water can damage crops, reduce fish populations and hurt farms downstream)

dynamics of habitat loss, decline of pollinators and other beneficial organisms and the disruption of natural processes vital for agriculture.

3.3.1 Habitat Loss Due to Urban Expansion and Land-Use Change

Urban expansion and associated land-use changes are among the primary drivers of habitat loss, leading to the fragmentation and degradation of ecosystems essential for supporting agricultural productivity. As urban areas expand, they encroach upon adjacent natural habitats, converting them into built-up environments or agricultural land. This transformation results in the loss of critical habitats such as forests, wetlands and grasslands, which harbour diverse species crucial for ecosystem functioning. The conversion of natural habitats into urban landscapes disrupts ecological processes, diminishes biodiversity and undermines the provision of ecosystem services necessary for agricultural production. Moreover, the loss of habitats due to urban expansion fragments ecosystems, leading to isolation of species populations and reduced gene flow, which can exacerbate biodiversity loss and compromise ecosystem resilience [63].

To mitigate habitat loss resulting from urban expansion, effective land-use planning strategies and conservation measures are imperative. Implementing green infrastructure initiatives, such as urban green spaces and wildlife corridors, can help preserve and restore habitats within urban environments. Additionally, promoting sustainable land-use practices, such as smart growth principles and land-use zoning regulations, can facilitate urban development while minimizing negative impacts on biodiversity and agricultural landscapes.

3.3.2 Decline of Pollinators and Other Beneficial Organisms

Pollinators and other beneficial organisms play a critical role in agricultural ecosystems by facilitating pollination, pest control and nutrient cycling. However, widespread habitat loss, pesticide use, pollution and climate change have contributed to the decline of pollinator populations and other beneficial organisms, posing significant challenges to agricultural sustainability. The loss of pollinators, including bees, butterflies and certain bird species, threatens the reproductive success of many crops dependent on insect-mediated pollination [64]. Without adequate pollination services, crop yields and quality may decline, leading to economic losses for farmers and food insecurity for communities reliant on agricultural production. Similarly, the decline of beneficial organisms such as predatory insects, soil microbes and vertebrate predators disrupts natural pest regulation mechanisms, increasing the vulnerability of crops to pest outbreaks and reducing the efficacy of integrated pest management strategies.

To address the decline of pollinators and beneficial organisms, conservation efforts focused on habitat restoration, reduced pesticide use and promotion of biodiversity-friendly agricultural practices are essential. Creating pollinator-friendly habitats within agricultural landscapes, such as flowering hedgerows and cover crops, can provide food and shelter for pollinators while enhancing ecosystem resilience. Furthermore, adopting agroecological approaches that prioritize ecological principles and biodiversity conservation can help restore balance to agricultural ecosystems and mitigate the negative impacts of pollinator decline [65].

3.3.3 Disruption of Natural Processes Essential for Agriculture

The disruption of natural processes essential for agriculture, such as nutrient cycling, soil formation and water regulation, poses significant challenges to sustainable food production. Human activities, including land conversion, deforestation and intensive agricultural practices, have altered ecological processes, leading to soil degradation, nutrient depletion and water scarcity. One of the critical consequences of disrupting natural processes is soil erosion, which accelerates due to the loss of vegetative cover and soil compaction resulting from land-use changes and agricultural activities. Soil erosion not only depletes soil fertility but also impairs water quality by contributing to sedimentation in water bodies, exacerbating the degradation of aquatic ecosystems. Moreover, the depletion of soil nutrients, particularly nitrogen, phosphorus and potassium, compromises soil health and reduces agricultural productivity. Intensive farming practices, such as monoculture and excessive fertilizer use, disrupt nutrient cycling processes, leading to nutrient imbalances, soil acidification and decreased soil organic matter content [66].

To mitigate the disruption of natural processes essential for agriculture, sustainable land management practices that promote soil conservation, biodiversity conservation and water resource management are paramount. Adopting conservation agriculture techniques, such as minimum tillage, cover cropping and agroforestry, can help prevent soil erosion, improve soil structure and enhance nutrient cycling processes. Additionally, implementing watershed management strategies, such as riparian buffers and soil conservation measures, can safeguard water resources and mitigate the adverse impacts of land-use changes on aquatic ecosystems. Addressing biodiversity loss and restoring ecosystem services are fundamental for enhancing agricultural sustainability and mitigating environmental degradation. By understanding the complex interconnections between urban expansion, habitat loss, pollinator decline and disruptions to natural processes, policymakers, practitioners and stakeholders can develop holistic strategies to promote resilience, conserve biodiversity and ensure the long-term viability of agricultural systems [67, 68].

3.4 Climate Change and Extreme Events

Climate change poses profound challenges to agricultural systems, altering temperature regimes, precipitation patterns and the frequency of extreme events, thereby exacerbating vulnerabilities and risks for farmers.

3.4.1 Impact of Rising Temperatures and Changing Precipitation Patterns

The phenomenon of climate change, primarily driven by anthropogenic activities, has resulted in a noticeable increase in global temperatures over the past century. These rising temperatures are profoundly impacting agricultural systems worldwide, with implications ranging from altered growing seasons to shifts in crop suitability and distribution patterns. One of the most evident impacts of rising temperatures on agriculture is the alteration of growing seasons. As temperatures increase, the timing of planting and harvesting cycles is being disrupted, leading to challenges in crop management and reduced yields. Additionally, higher temperatures can accelerate the maturation of crops, affecting their quality and nutritional value. In regions where temperature increases exceed the tolerance limits of certain crops, there is a risk of decreased productivity or even crop failure [69]. Changing precipitation patterns further exacerbate the challenges faced by agricultural systems. Some regions experience shifts in rainfall distribution, with increased variability and intensity of precipitation events. This variability can disrupt planting schedules, promote soil erosion and lead to waterlogging or drought stress, depending on the timing and magnitude of precipitation. Moreover, changes in precipitation patterns can affect soil moisture levels, impacting crop growth and nutrient uptake.

Furthermore, altered precipitation patterns contribute to the phenomenon of water scarcity, particularly in regions where water resources are already limited. Inadequate access to water for irrigation purposes can severely hinder agricultural productivity, exacerbating food insecurity and economic hardship for farming communities. To mitigate the adverse effects of rising temperatures and changing precipitation patterns on agriculture, adaptation strategies are crucial [70]. These may include the development and adoption of heat-tolerant crop varieties, implementation of water-saving irrigation techniques and promotion of climate-smart agricultural practices such as agroforestry and conservation agriculture. Additionally, policymakers must prioritize investments in climate-resilient infrastructure and enhance early warning systems to help farmers anticipate and adapt to changing climatic conditions [71].

3.4.2 Increased Frequency and Intensity of Droughts and Floods

Climate change is amplifying the frequency and intensity of extreme weather events, including droughts and floods, posing significant challenges to agricultural production and food security. Droughts, characterized by prolonged periods of deficient precipitation and water scarcity, can have devastating effects on crop yields, livestock productivity and rural livelihoods. During droughts, crops are particularly vulnerable to moisture stress, leading to wilting, stunted growth and reduced photosynthetic activity. In severe cases, prolonged droughts can result in complete crop failure, causing substantial economic losses for farmers and threatening food availability for communities. Moreover, water scarcity exacerbates competition for limited resources among agricultural, industrial and domestic users, intensifying conflicts over water allocation and management.

Conversely, climate change also contributes to the intensification of flooding events, characterized by excessive rainfall or rapid snowmelt, which can inundate agricultural fields, destroy crops and erode soil fertility. Floods not only lead to direct physical damage to crops and infrastructure but also increase the risk of waterborne diseases, soil erosion and sedimentation of water bodies. In regions prone to recurrent flooding, farmers face challenges in rebuilding infrastructure, restoring soil fertility and resuming agricultural activities in the aftermath of extreme events [72]. Addressing the increased frequency and intensity of droughts and floods requires a multifaceted approach that integrates disaster risk reduction, water management and climate adaptation measures. Investing in drought-resistant crop varieties, improving water harvesting and storage infrastructure and implementing sustainable land management practices can enhance the resilience of agricultural systems to extreme weather events. Moreover, strengthening early warning systems, promoting insurance schemes and providing social safety nets can support vulnerable farming communities in coping with the impacts of droughts and floods.

3.4.3 Vulnerability of Agricultural Systems to Climate Shocks

Agricultural systems are inherently vulnerable to climate shocks, which encompass a wide range of extreme weather events, including heatwaves, storms, frosts and pest outbreaks. Climate shocks disrupt the normal functioning of agricultural production systems, causing significant economic losses, food insecurity and social upheaval. One of the primary factors contributing to the vulnerability of agricultural systems to climate shocks is their dependence on environmental conditions and natural resources. Unlike other sectors, such as manufacturing or services, agriculture is heavily reliant on climatic factors such as temperature, rainfall and sunlight for crop growth and livestock rearing. Consequently, any deviations from the optimal environmental conditions can have profound implications for agricultural productivity and livelihoods [73].

Moreover, the interconnectedness of agricultural systems with other sectors of the economy amplifies the ripple effects of climate shocks. For example, disruptions in agricultural production can lead to food shortages, price volatility and inflation, impacting food security and livelihoods across urban and rural areas. Furthermore, climate shocks can exacerbate existing vulnerabilities, such as poverty, land degradation and limited access to markets and financial resources, disproportionately affecting marginalized communities and smallholder farmers. In addition to environmental and socioeconomic factors, the institutional capacity to respond to climate shocks plays a crucial role in determining the resilience of agricultural systems [74]. Effective governance structures, robust policies and institutional mechanisms for disaster risk management and adaptation are essential for reducing the vulnerability of farming communities to climate-related hazards. Strengthening extension services, promoting knowledge exchange and capacity-building initiatives and facilitating access to climate information and technology can empower farmers to adopt climate-resilient practices and cope with the impacts of climate shocks. Addressing the vulnerability of agricultural systems to climate shocks requires coordinated efforts at the local, national and international levels, involving policymakers, researchers, civil society organizations and the private sector. By enhancing the resilience of agricultural systems, promoting sustainable practices and investing in climate adaptation and disaster risk reduction measures, societies can safeguard food security, protect livelihoods and build a more sustainable and resilient future for generations to come [75, 76].

4 Policy and Governance

Government policies play a crucial role in shaping the urban landscape and influencing environmental practices. This section identifies the significance of government intervention, the challenges faced in implementing sustainable urban policies and strategies required for effective governance to mitigate environmental degradation.

4.1 Role of Government Policies in Addressing Urban Sustainability

Government policies play a pivotal role in addressing urban sustainability by providing a regulatory framework, incentives and directives that guide urban development towards more sustainable practices. The primary objectives of these policies are to promote resource efficiency, reduce environmental impact, enhance resilience to climate change and improve the overall quality of urban life. Some aspects of government policies in this context include:

- ***Regulatory Framework***: Governments enact laws and regulations that set standards for urban development, land use, building codes and environmental protection. These regulations ensure that new developments comply with sustainability criteria such as energy efficiency, waste management and green infrastructure.
- ***Incentives and Subsidies***: Governments offer incentives such as tax breaks, grants and subsidies to encourage sustainable practices among businesses, developers and residents. This can include incentives for adopting renewable energy, implementing green building designs, or using public transportation [77].
- ***Integrated Planning***: Urban sustainability policies promote integrated planning approaches that consider social, economic and environmental factors holistically. This involves coordination between different government departments, stakeholders and communities to achieve sustainable development goals.
- ***Public Engagement***: Effective policies also prioritize public engagement and participation in decision-making processes. This ensures that urban sustainability policies align with community needs and values, enhancing their acceptance and effectiveness.
- ***Monitoring and Reporting***: Governments establish monitoring mechanisms to track progress towards sustainability goals and report on key indicators such as greenhouse gas emissions, air and water quality and urban biodiversity. This enables evidence-based policy adjustments and accountability [78, 79].

4.2 Challenges in Implementing Sustainable Urban Policies

Despite the importance of sustainable urban policies, several challenges hinder their effective implementation:

- ***Complex Interdependencies***: Urban systems are characterized by complex interdependencies between various sectors (e.g., transportation, energy, waste management). Coordinating policies across these sectors requires overcoming institutional and administrative barriers.
- ***Financial Constraints***: Implementing sustainable practices often requires substantial upfront investments, which may strain public budgets. Securing long-term funding and overcoming cost barriers is essential for successful implementation.

- ***Policy Fragmentation***: Inconsistent policies across different levels of government (local, regional, national) or between sectors can create confusion and hinder coherent implementation [80]. Achieving policy coherence and coordination is crucial for effective sustainability outcomes.
- ***Lack of Awareness and Capacity***: Many stakeholders, including policymakers, developers and the public, may lack awareness or capacity to adopt sustainable practices. Education and capacity-building efforts are essential to overcome this barrier.
- ***Resistance to Change***: Resistance from vested interests, existing economic models, or established practices can impede the adoption of innovative sustainable policies. Overcoming inertia and fostering behavioural change is a significant challenge.

4.3 Strategies for Effective Governance

To enhance the effectiveness of governance in promoting urban sustainability, several strategies can be employed:

- ***Integrated and Coordinated Approaches***: Governments should adopt integrated planning approaches that involve multiple stakeholders and sectors. This requires breaking down silos between departments and fostering collaboration.
- ***Capacity Building and Education***: Investing in capacity building and raising awareness among policymakers, professionals and the public about the benefits and methods of sustainable urban development [81].
- ***Policy Innovation and Experimentation***: Governments should encourage policy experimentation and innovation by piloting new approaches on a smaller scale before scaling up successful initiatives.
- ***Public Participation and Transparency***: Engaging citizens and stakeholders in decision-making processes through participatory mechanisms such as public consultations, community workshops and online platforms.
- ***Incentives and Regulations***: Governments should use a mix of incentives (e.g., subsidies, tax breaks) and regulations (e.g., zoning laws, building codes) to drive sustainable practices in urban development.
- ***Monitoring and Evaluation***: Establishing robust monitoring and evaluation systems to track progress, identify bottlenecks and adjust policies accordingly based on empirical evidence [82].

Effective governance in promoting urban sustainability requires a comprehensive and coordinated approach that integrates policies, engages stakeholders, fosters innovation and ensures accountability. Addressing challenges such as policy fragmentation, financial constraints and resistance to change is essential for achieving sustainable urban development goals [83].

5 Technological Solutions

In addressing the multifaceted challenges posed by urban sustainability and environmental degradation on agricultural production, technological solutions play a pivotal role. This section Discusses various technological innovations aimed at ameliorating these challenges, including advancements in urban agriculture techniques, technological approaches to mitigate environmental degradation and the integration of technology for sustainable urban environments.

5.1 Innovations in Urban Agriculture Techniques

Urban agriculture has emerged as a promising avenue to enhance food security, mitigate environmental impacts and foster community resilience in urban settings. Technological innovations in this realm encompass a wide array of approaches, ranging from vertical farming and hydroponics to aquaponics and rooftop gardening [84]. Vertical farming, for instance, involves cultivating crops in vertically stacked layers, often within controlled indoor environments. This method optimizes land use efficiency, reduces water consumption and minimizes the need for chemical pesticides by employing hydroponic or aeroponic systems. Additionally, advancements in LED lighting technology enable precise control over light spectra, duration and intensity, thus enhancing crop yields and quality.

Similarly, hydroponic and aquaponic systems leverage recirculating nutrient solutions and symbiotic relationships between fish and plants to cultivate crops in soil-less environments. These techniques not only conserve water but also mitigate nutrient runoff and pollution associated with conventional agricultural practices [85]. Rooftop gardening represents another innovative approach to urban agriculture, utilizing underutilized urban spaces to cultivate fruits, vegetables and herbs. By incorporating green roofs into urban infrastructure, cities can mitigate heat island effects, improve air quality and enhance biodiversity while simultaneously enhancing food sovereignty and community cohesion.

5.2 Technological Approaches to Mitigate Environmental Degradation

Environmental degradation resulting from urbanization and intensive agricultural practices poses significant challenges to sustainability and ecosystem health. Technological interventions offer viable solutions to mitigate these adverse impacts, encompassing soil remediation, waste management and pollution control measures. Soil remediation technologies, such as phytoremediation and biochar application, facilitate the restoration of degraded soils by removing contaminants and enhancing soil fertility. Phytoremediation harnesses the natural detoxification abilities of plants to absorb, metabolize, or sequester pollutants, thereby detoxifying contaminated soils. Biochar, a carbon-rich material derived from biomass pyrolysis, improves soil structure, water retention and nutrient cycling while sequestering carbon dioxide from the atmosphere [86, 87].

Effective waste management strategies are imperative for mitigating environmental pollution and conserving natural resources in urban environments. Technological innovations in waste-to-energy conversion, recycling and composting enable the valorisation of organic waste streams while minimizing landfill emissions and resource depletion. Anaerobic digestion, for example, converts organic waste into biogas, a renewable energy source and nutrient-rich digestate, which can be utilized as a biofertilizer. Furthermore, pollution control technologies, such as electrostatic precipitators, wet scrubbers and catalytic converters, mitigate air and water pollution emissions from industrial activities, transportation and agricultural runoff. These systems remove particulate matter, volatile organic compounds and hazardous pollutants from exhaust gases and wastewater streams, thereby safeguarding human health and environmental quality [88].

5.3 Integration of Technology for Sustainable Urban Environments

The integration of technology holds immense potential to foster sustainable urban environments by optimizing resource utilization, enhancing resilience and promoting equitable development. Smart city initiatives leverage data-driven technologies, Internet of Things (IoT) sensors and artificial intelligence (AI) algorithms to optimize urban infrastructure and services, including energy distribution, transportation networks and waste management systems. For instance, smart grid technologies enable dynamic energy management, demand response and distributed renewable energy generation, reducing carbon emissions and enhancing grid resilience. Advanced transportation systems, such as electric vehicles (EVs) and autonomous vehicles, mitigate air pollution and congestion while promoting sustainable mobility solutions.

Moreover, digital platforms and citizen engagement tools empower communities to actively participate in decision-making processes, fostering social inclusivity and civic awareness. By harnessing the collective intelligence of residents, cities can co-create innovative solutions to address local challenges and enhance quality of life [89]. Technological solutions offer valuable tools for addressing the complex interplay between urban sustainability, environmental degradation and agricultural production. By embracing innovation and fostering interdisciplinary collaboration, cities can pave the way towards a more resilient, equitable and environmentally sustainable future [90].

6 Community Engagement and Education

In addressing the intricate interplay between urban sustainability and agricultural production, community engagement and education emerge as pivotal factors. This section identifies the significance of community involvement in sustainable urban practices, the role of educational initiatives in fostering urban sustainability and strategies for promoting awareness and behavioural change among urban populations.

6.1 Importance of Community Involvement in Sustainable Urban Practices

Community involvement stands as a cornerstone of sustainable urban development. Engaging local residents, businesses and organizations in decision-making processes fosters a sense of ownership and accountability towards environmental stewardship. In the context of agricultural production, urban communities play a crucial role in advocating for sustainable farming practices, supporting local food systems and mitigating environmental impacts [91].

Effective community involvement entails inclusive participation mechanisms that empower diverse stakeholders to contribute their perspectives and expertise. Collaborative platforms such as community gardens, farmers' markets and urban agriculture initiatives not only promote sustainable food production but also cultivate a sense of community cohesion and resilience. Moreover, community engagement serves as a catalyst for innovation and knowledge sharing, facilitating the exchange of best practices and the co-creation of sustainable solutions tailored to local contexts [92]. By harnessing the collective wisdom and resources of urban residents, municipalities can address complex challenges such as food insecurity, land degradation and climate change resilience in a holistic manner.

6.2 Educational Initiatives for Urban Sustainability

Educational initiatives play a pivotal role in nurturing a culture of sustainability within urban communities. By equipping individuals with the knowledge, skills and values necessary to make informed choices, education serves as a catalyst for transformative change towards more sustainable lifestyles and consumption patterns.

In the context of urban agriculture, educational programs play a dual role in fostering both environmental literacy and agricultural proficiency. By integrating concepts of ecology, food systems and sustainable agriculture into formal and informal education curricula, schools, universities and community organizations can cultivate a new generation of environmentally conscious citizens and skilled urban farmers. Furthermore, hands-on learning experiences such as farm-to-school programs, experiential workshops and vocational training opportunities provide valuable opportunities for individuals to engage directly with food production processes, deepen their understanding of ecological principles and develop practical skills for sustainable living [93, 94].

6.3 Promoting Awareness and Behavior Change

Promoting awareness and behaviour change is essential for catalysing sustainable urban transformations. Effective communication strategies, targeted outreach campaigns and behaviour change interventions are instrumental in raising awareness about the environmental impacts of urbanization and fostering a sense of collective responsibility among citizens. One key strategy for promoting behaviour change is through social marketing techniques that leverage psychological insights and behavioural science principles to influence attitudes, beliefs and actions towards sustainability. By framing sustainability messages in a relatable, actionable and culturally sensitive manner, communicators can inspire individuals to adopt pro-environmental behaviours and support sustainable initiatives [95].

Additionally, community-based approaches such as peer-to-peer learning networks, neighbourhood mobilization campaigns and citizen science projects empower individuals to take ownership of sustainability issues and drive grassroots change within their communities. By fostering a sense of agency and collective efficacy, these initiatives amplify the impact of individual actions and catalyse broader societal shifts towards a more sustainable future [96]. Community engagement and education are indispensable pillars of urban sustainability and environmental stewardship. By fostering inclusive participation, promoting lifelong learning and empowering citizens to take action, municipalities can harness the collective power of communities to address pressing challenges such as agricultural production, resource depletion and climate change adaptation in urban settings [97].

7 Economic Perspectives

In examining the economic dimensions of urban sustainability and its impact on agricultural production, several critical aspects merit detailed scrutiny. This section examines the economic implications of unsustainable urban practices on agriculture, conducts a cost–benefit analysis of sustainable urban solutions and explores market incentives conducive to sustainable urban development.

7.1 Economic Impacts of Unsustainable Urban Practices on Agriculture

Urbanization, characterized by the rapid expansion of cities and the associated increase in population density, has profound implications for agriculture and food production. However, the economic impacts of unsustainable urban practices on agriculture are multifaceted and complex, encompassing various factors such as land use change, resource depletion and market dynamics. This section identifies the relationship between urbanization and its detrimental effects on agricultural production, examining both the theoretical underpinnings and practical examples [98].

To comprehend the economic impacts of unsustainable urban practices on agriculture, it is imperative to employ a theoretical framework that elucidates the interplay between urbanization and agricultural dynamics. One such framework is the Land Rent Theory, which posits that the value of land is determined by its productivity and proximity to urban centres. As urban areas expand, agricultural land becomes increasingly scarce and valuable, leading to escalating land rents and land use conversion. Moreover, the von Thünen model provides insights into the spatial organization of agricultural activities in relation to urban markets. According to this model, agricultural land use patterns are shaped by transportation costs and market demand, with intensive agriculture closer to urban centres and extensive agriculture farther away. However, unsustainable urban practices such as indiscriminate land development and fragmentation can disrupt this spatial organization, resulting in suboptimal land use patterns and reduced agricultural productivity.

The economic impacts of unsustainable urban practices on agriculture manifest through several mechanisms, each influencing agricultural production, profitability and sustainability. One such mechanism is the loss of prime agricultural land to urban expansion, which diminishes the agricultural land base and increases competition for limited land resources. This, in turn, drives up land prices and rents, rendering farming economically unviable for smallholders and exacerbating inequities in land access. Additionally, unsustainable urban practices contribute to environmental degradation, such as soil erosion, water pollution and habitat loss, which compromise agricultural productivity and necessitate costly remediation measures. Furthermore, the proliferation of urban sprawl and infrastructure development encroaches upon agricultural areas, fragmenting landscapes and disrupting traditional farming practices [99].

To quantify the economic impacts of unsustainable urban practices on agriculture, one can employ a simple formula that accounts for changes in land value, agricultural productivity and externalities such as pollution and habitat destruction.

$$Economic\ Impact = \Delta\ Land\ Value + \Delta\ Agricultural\ Productivity - Cost\ of\ Externalities$$

For example, consider a scenario where a peri-urban agricultural region experiences rapid urbanization, leading to a 20% decrease in available agricultural land, a 10% decline in crop yields due to pollution and an additional expenditure of $50,000 for environmental remediation efforts. Using the formula:

$$\text{Economic Impact} = (-0.20 \times \text{Initial Land Value}) + (-0.10 \times \text{Initial Agricultural Productivity}) - \$50,000$$

By plugging in the respective values for land value and agricultural productivity, one can quantify the economic impact of unsustainable urban practices on agriculture in monetary terms.

7.2 Cost–Benefit Analysis of Sustainable Urban Solutions

Urban sustainability initiatives aim to mitigate environmental degradation and enhance the overall quality of life in urban areas. However, implementing such solutions often incurs costs and assessing their effectiveness requires a comprehensive cost–benefit analysis (CBA).

A CBA is a systematic approach used to evaluate the economic feasibility of a project or policy by comparing its costs with its benefits. In the context of sustainable urban solutions, this involves quantifying the costs associated with implementing initiatives such as renewable energy projects, green infrastructure development, or waste management programs and comparing them with the anticipated benefits in terms of environmental improvement, economic gains and social well-being. The general formula for conducting a CBA involves calculating the net present value (NPV) of the costs and benefits associated with the proposed intervention over a specified time horizon [100]. The NPV is calculated by discounting all future costs and benefits to their present values using an appropriate discount rate. The formula for NPV is as follows:

$$NPV = \sum_{t=1}^{n} \frac{B_t - C_y}{(1+r)^t}$$

where:

- NPV = Net Present Value
- Bt = Benefits in period t
- Ct = Costs in period t
- r = Discount rate
- n = Number of periods

Example: Consider a city planning to implement a sustainable transportation project aimed at reducing carbon emissions and traffic congestion. The project involves the construction of dedicated bike lanes and pedestrian walkways, as well as the introduction of a bike-sharing program. The estimated costs and benefits of the project over a 10-year period are as follows:

- Initial Investment (Year 0): $1,000,000
- Annual Operational Costs: $50,000

- Annual Benefits (e.g., reduced healthcare costs, increased productivity): $200,000

Assuming a discount rate of 5%, we can calculate the NPV of the project using the formula provided.

$$NVP = \frac{200{,}000 - 50{,}000}{(1+0.05)^1} + \frac{200{,}000 - 50{,}000}{(1+0.05)^2} + \cdots + \frac{200{,}000 - 50{,}000}{(1+0.05)^{10}} - 1{,}000{,}000$$

$$NVP = \frac{150{,}000}{1.05^1} + \frac{150,000}{1.05^2} + \cdots + \frac{150{,}000}{1.05^{10}} - 1{,}000{,}000$$

$$NVP \approx \frac{150{,}000}{1.05} + \frac{150{,}000}{1.1025} + \cdots + \frac{150{,}000}{1.6289} - 1{,}000{,}000$$

$$NVP \approx 142{,}857.14 + 136{,}612.02 + \cdots + 55{,}914.67 - 1{,}000{,}000$$

$$NVP \approx 488{,}054.41 - 1{,}000{,}000$$

$$NVP \approx -511{,}945.59$$

In this example, the negative NPV indicates that the present value of the project's costs outweighs the present value of its benefits. Therefore, based on this analysis, the sustainable transportation project may not be economically viable given the specified parameters.

A thorough CBA is essential for evaluating the economic feasibility and potential impacts of sustainable urban solutions. By quantifying both the costs and benefits associated with these initiatives, policymakers and stakeholders can make informed decisions regarding resource allocation and prioritize interventions that maximize societal welfare while minimizing financial burdens.

7.3 Market Incentives for Sustainable Urban Development

Market incentives play a pivotal role in driving sustainable urban development initiatives. These incentives, often leveraged by governments, businesses and non-profit organizations, aim to align economic interests with environmental and social objectives, thereby fostering sustainable practices within urban areas. This section discusses the various market incentives that encourage sustainable urban development and their impact on agricultural production within the urban context.

7.3.1 Carbon Pricing Mechanisms

Carbon pricing mechanisms, such as carbon taxes and cap-and-trade systems, are prominent market-based instruments aimed at internalizing the external costs of carbon emissions. By assigning a price to carbon, these mechanisms incentivize businesses and individuals to reduce their carbon footprint. In the context of urban sustainability, carbon pricing can drive investments in renewable energy, energy-efficient infrastructure and low-carbon transportation systems. Such investments not only mitigate climate change but also contribute to reducing environmental degradation associated with agricultural production. For instance, the adoption of clean energy technologies in urban areas can reduce the demand for fossil fuels, thereby curbing emissions from agricultural machinery and transportation [101].

7.3.2 Green Subsidies and Tax Incentives

Governments often provide green subsidies and tax incentives to promote sustainable practices in urban development. These incentives may include grants, subsidies, or tax breaks for businesses and individuals adopting environmentally friendly technologies and practices. In the context of agriculture, green subsidies can encourage urban farmers to implement sustainable farming techniques, such as agroforestry, organic farming and water-efficient irrigation systems. By subsidizing the adoption of these practices, governments not only support local food production but also mitigate the environmental impacts associated with conventional agricultural methods, such as chemical runoff and soil degradation.

7.3.3 Tradable Development Rights

Tradable development rights (TDRs) represent another market-based approach to incentivize sustainable urban development. TDRs allow property owners to transfer development rights from one parcel of land to another, typically from rural or environmentally sensitive areas to urban centres. By incentivizing densification and preserving green spaces, TDR programs promote compact, walkable urban environments while protecting agricultural land and natural habitats. From an agricultural standpoint, TDRs can help mitigate urban sprawl and land-use conflicts by preserving farmland on the urban fringe. This not only maintains local food production capacity but also safeguards ecosystem services provided by agricultural landscapes, such as pollination and water filtration [102].

7.3.4 Eco-Certification and Labelling Programs

Eco-certification and labelling programs provide market incentives for sustainable production practices by certifying products that meet specific environmental and social criteria. In the urban context, eco-certification schemes for agricultural products can enhance consumer awareness and preference for sustainably produced food. Farmers adhering to certified standards may gain access to premium markets and price premiums, incentivizing the adoption of sustainable farming practices [103]. Moreover, eco-certification programs often require transparent supply chain management and traceability, fostering accountability and environmental stewardship throughout the agricultural value chain.

7.3.5 Green Procurement Policies

Public and private sector entities can leverage their purchasing power to drive demand for sustainable goods and services through green procurement policies. By prioritizing environmentally preferable products and suppliers, governments, corporations and institutions can create market opportunities for sustainable urban solutions. In the context of agriculture, green procurement policies can support local farmers and producers who adhere to sustainable farming practices. By sourcing food locally and sustainably, urban entities can reduce the environmental footprint of their food supply chains, minimize transportation-related emissions and enhance food security and resilience [104].

7.3.6 Social Impact Investing

Social impact investing involves allocating capital to projects and enterprises that generate positive social and environmental outcomes alongside financial returns. In the realm of sustainable urban development, social impact investors may support initiatives aimed at enhancing food security, promoting urban agriculture and fostering community resilience. By providing funding and expertise to sustainable agriculture ventures in urban areas, social impact investors can catalyse innovation and scale solutions that address environmental degradation while meeting the food needs of urban populations. This form of investment not only stimulates economic growth but also fosters inclusive and sustainable development [105].

7.3.7 Collaboration and Knowledge Sharing

Collaboration and knowledge sharing among stakeholders are essential for maximizing the effectiveness of market incentives for sustainable urban development. Governments, businesses, non-profit organizations, academia and communities must collaborate to design and implement incentive mechanisms that address the multifaceted challenges of urban sustainability and environmental degradation. By sharing best practices, data and resources, stakeholders can identify synergies, overcome barriers and leverage collective expertise to drive positive change [106]. Additionally, fostering partnerships between urban and rural stakeholders facilitates the integration of agricultural considerations into urban sustainability strategies, promoting holistic approaches to food systems resilience and environmental stewardship.

Market incentives play a crucial role in promoting sustainable urban development and mitigating environmental degradation associated with agricultural production. Through carbon pricing mechanisms, green subsidies, tradable development rights, eco-certification programs, green procurement policies, social impact investing and collaborative initiatives, stakeholders can harness market forces to incentivize sustainable practices, enhance food security and foster resilience in urban environments. However, effective implementation and monitoring are essential to ensure that market incentives align with broader sustainability goals and address the unique challenges of urbanization and agricultural production [107].

8 Conclusion

The ever-expanding urban footprint presents a significant challenge to sustainable agricultural production. Land conversion, water resource competition and air pollution from cities directly threaten agricultural productivity. The Food and Agriculture Organization (FAO) estimates that by 2050, the world will need to produce 70% more food to feed a growing population. Reconciling urban development with agricultural sustainability demands a multi-pronged approach. Policy and governance play a critical role in promoting sustainable urban practices. Incentives for urban agriculture, stricter regulations on pollution control and investments in green infrastructure are crucial. Technological advancements in areas like vertical farming and precision agriculture offer promising solutions for boosting yields within limited space. However, technology alone is not enough. Community engagement and education are essential for fostering a culture of sustainability. Raising public awareness about the interconnectedness of urban and rural systems can drive behaviour change towards responsible consumption and waste reduction. A strong economic case can be made for sustainable urban development. The economic costs of environmental degradation on agriculture are significant, impacting food security and livelihoods. Investing in sustainable practices, on the other hand, generates long-term economic benefits, fostering healthier communities, creating new green jobs and ensuring a reliable food supply for future generations. By adopting a holistic approach that integrates policy, technology, community engagement and economic incentives, we can create a future where thriving cities coexist with productive and resilient agricultural landscapes.

References

1. Urbanization IM (2003) Urban environment and land use: challenges and opportunities. In: Asia and the Pacific Forum for Environment and Development, Expert Meeting, vol. 23, 2003, January
2. Daramola A, Ibem EO (2010) Urban environmental problems in Nigeria: implications for sustainable development. J Sustain Dev Afr 12(1):124–145
3. Verma P, Raghubanshi AS (2018) Urban sustainability indicators: challenges and opportunities. Ecol Indic 93:282–291. https://doi.org/10.1016/j.ecolind.2018.05.007
4. Specht K, Siebert R, Hartmann I, Freisinger UB, Sawicka M, Werner A, Thomaier S, Henckel D, Walk H, Dierich A (2014) Urban agriculture of the future: An overview of sustainability aspects of food production in and on buildings. Agric Hum Values 31(1):33–51. https://doi.org/10.1007/s10460-013-9448-4
5. Langemeyer J, Madrid-Lopez C, Mendoza Beltran A, Villalba Mendez G (2021) Urban agriculture—A necessary pathway towards urban resilience and global sustainability? Landsc Urban Plan 210:104055. https://doi.org/10.1016/j.landurbplan.2021.104055
6. Finco A, Nijkamp P (2001) Pathways to urban sustainability. J Environ Policy Plan 3(4):289–302. https://doi.org/10.1002/jepp.94
7. Capps KA, Bentsen CN, Ramírez A (2016) Poverty, urbanization, and environmental degradation: urban streams in the developing world. Freshw Sci 35(1):429–435. https://doi.org/10.1086/684945
8. Arora NK, Fatima T, Mishra I, Verma M, Mishra J, Mishra V (2018) Environmental sustainability: challenges and viable solutions. Environ Sustain 1(4):309–340. https://doi.org/10.1007/s42398-018-00038-w
9. Kumar N, Kushwaha RR, Meena NR, Mishra H, Yadav APS (2023) A study on costs and returns of paddy cultivation in Ambedkar Nagar district of Uttar Pradesh. Int J Stat Appl Math SP 8(3):107–111
10. Lu Y, Jenkins A, Ferrier RC, Bailey M, Gordon IJ, Song S, Huang J, Jia S, Zhang F, Liu X, Feng Z, Zhang Z (2015) Addressing China's grand challenge of achieving food security while ensuring environmental sustainability. Sci Adv 1(1):e1400039. https://doi.org/10.1126/sciadv.1400039
11. Wadumestrige Dona CG, Mohan G, Fukushi K (2021) Promoting urban agriculture and its opportunities and challenges—A global review. Sustainability 13(17):9609. https://doi.org/10.3390/su13179609
12. Drechsel P, Dongus S (2010) Dynamics and sustainability of urban agriculture: Examples from Sub-Saharan Africa. Sustain Sci 5(1):69–78. https://doi.org/10.1007/s11625-009-0097-x
13. Ameen RFM, Mourshed M (2017) Urban environmental challenges in developing countries—A stakeholder perspective. Habitat Int 64:1–10. https://doi.org/10.1016/j.habitatint.2017.04.002
14. Yu J, Wu J (2018) The sustainability of agricultural development in China: The agriculture-environment nexus. Sustainability 10(6):1776. https://doi.org/10.3390/su10061776
15. Hossain A, Krupnik TJ, Timsina J, Mahboob MG, Chaki AK, Farooq M, Hasanuzzaman M (2020) Agricultural land degradation: processes and problems undermining future food security. In: Environment, Climate, Plant and Vegetation Growth; Springer International Publishing, pp 17–61
16. Hallett S, Hoagland L, Toner E (2016) Urban agriculture: environmental, economic, and social perspectives. Hortic Rev 44(44):65–120
17. Nwokoro CV, Chima FO (2017) Impact of environmental degradation on agricultural production and poverty in Rural Nigeria. Am Int J Contemp Res 7(2):6–14
18. Mishra H (2024) Counting the costs: Understanding agricultural economics and the concept of costs. In: Sharma K, Singh P, Singh S, Yadav A, Awasthi V (Eds) Text Book on Practices for Agricultural Sustainability. Golden Leaf Publishers, pp 198–210

19. Li W, Wang D, Li Y, Zhu Y, Wang J, Ma J (2020) A multi-faceted, location-specific assessment of land degradation threats to Periurban agriculture at a traditional grain base in Northeastern China. J Environ Manag 271:111000. https://doi.org/10.1016/j.jenvman.2020.111000
20. Das J, Jha S, Goyal MK, Surampalli RY (2020) Challenges of sustainability in agricultural management. Sustain Fundam Appl 339–356
21. Drechsel P, Cofie O, Niang S (2008) Sustainability and resilience of the urban agricultural phenomenon in Africa. Conserv Land Prot Water 6:120
22. Aubry C, Ramamonjisoa J, Dabat M-H, Rakotoarisoa J, Rakotondraibe J, Rabeharisoa L (2012) Urban agriculture and land use in cities: An approach with the multi-functionality and sustainability concepts in the case of Antananarivo (Madagascar). Land Use Policy 29(2):429–439. https://doi.org/10.1016/j.landusepol.2011.08.009
23. Titilola SO, Jeje LK (2008) Environmental degradation and its implications for agricultural and rural development: The issue of land erosion. J. Sustain. Dev. Afr. 10(2):116–146
24. Zhao J, Luo Q, Deng H, Yan Y (2008) Opportunities and challenges of sustainable agricultural development in China. Philos Trans R Soc Lond B Biol Sci 363(1492):893–904. https://doi.org/10.1098/rstb.2007.2190
25. Zellner ML, Theis TL, Karunanithi AT, Garmestani AS, Cabezas H (2008) A new framework for urban sustainability assessments: Linking complexity, information and policy. Comput Environ Urban Syst 32(6):474–488. https://doi.org/10.1016/j.compenvurbsys.2008.08.003
26. Cahya DL (2016) Analysis of urban agriculture sustainability in metropolitan Jakarta (Case Study: Urban Agriculture in Duri Kosambi). Procedia Soc Behav Sci 227:95–100. https://doi.org/10.1016/j.sbspro.2016.06.048
27. Mishra H (2024) Molecular Technology in Agriculture. In: Patel DK, Dwivedi S, Dubey A, Kumar A, Ali W (Eds) Agricultural Transformation in the Modern Era. Elite Publishing House, pp 204–226
28. Azunre GA, Amponsah O, Peprah C, Takyi SA, Braimah I (2019) A review of the role of urban agriculture in the sustainable city discourse. Cities 93:104–119. https://doi.org/10.1016/j.cities.2019.04.006
29. Agudelo-Vera CM, Mels AR, Keesman KJ, Rijnaarts HHM (2011) Resource management as a key factor for sustainable urban planning. J Environ Manag 92(10):2295–2303. https://doi.org/10.1016/j.jenvman.2011.05.016
30. Nugent RA (1999) Measuring the sustainability of urban agriculture. For hunger-proof cities Sustain Urban Food Syst 95–99
31. Simon D (2008) Urban environments: Issues on the Periurban fringe. Annu Rev Environ Resour 33(1):167–185. https://doi.org/10.1146/annurev.environ.33.021407.093240
32. Næss P (2001) Urban planning and sustainable development. Eur Plan Stud 9(4):503–524. https://doi.org/10.1080/09654310120049871
33. Hiremath RB, Balachandra P, Kumar B, Bansode SS, Murali J (2013) Indicator-based urban sustainability—A review. Energy Sustain Dev 17(6):555–563. https://doi.org/10.1016/j.esd.2013.08.004
34. Feola G, Suzunaga J, Soler J, Wilson A (2020) Periurban agriculture as quiet sustainability: Challenging the urban development discourse in Sogamoso Colombia. J Rural Stud 80:1–12. https://doi.org/10.1016/j.jrurstud.2020.04.032
35. Kumar JR, Natasha B, Suraj KC, Kumar SA, Manahar K (2019) Rooftop farming: An alternative to conventional farming for urban sustainability. Malays J Sustain Agric 3(1):39–43. https://doi.org/10.26480/mjsa.01.2019.39.43
36. Wang G, Shi R, Mi L, Hu J (2022) Agricultural Eco-efficiency: Challenges and progress. Sustainability14(3):1051. https://doi.org/10.3390/su14031051
37. Panagopoulos T, González Duque JA, Bostenaru Dan M (2016) Urban planning with respect to environmental quality and human well-being. Environ Pollut 208(A):137–144. https://doi.org/10.1016/j.envpol.2015.07.038
38. Ramaiah M, Avtar R (2019) Urban green spaces and their need in cities of rapidly urbanizing India: A review. Urban Sci 3(3):94. https://doi.org/10.3390/urbansci3030094

39. Tiwari AK, Mishra H, Nishad DC, Pandey A (2023) Sustainable water management in agriculture: Irrigation techniques and water conservation. Dr Ajay B Jadhao 53
40. Drescher A (2001) The integration of urban agriculture into urban planning–An analysis of the current status and constraints. Annot Bibliogr Urban Agric. ETC-RUAF/CTA
41. Vlek PL, Khamzina A, Azadi H, Bhaduri A, Bharati L, Braimoh A, Martius C, Sunderland T, Taheri F (2017) Trade-offs in Multi-purpose land use under land degradation. Sustainability 9(12):2196. https://doi.org/10.3390/su9122196
42. Sturiale L, Scuderi A (2019) The role of green infrastructures in urban planning for climate change adaptation. Climate 7(10):119. https://doi.org/10.3390/cli7100119
43. Regmi PP, Weber KE (2000) Problems to agricultural sustainability in developing countries and a potential solution: Diversity. Int J Soc Econ 27(7/8/9/10):788–801. https://doi.org/10.1108/03068290010335226
44. Indraprahasta GS (2013) The potential of urban agriculture development in Jakarta. Procedia Environ Sci 17:11–19. https://doi.org/10.1016/j.proenv.2013.02.006
45. Basiago AD (1998) Economic, social, and environmental sustainability in development theory and urban planning practice. Environmentalist 19(2):145–161. https://doi.org/10.1023/A:1006697118620
46. Asomani-Boateng R, Haight M (1999) Reusing organic solid waste in urban farming in African cities: A challenge for urban planners. Third World Plann Rev 21(4):411. https://doi.org/10.3828/twpr.21.4.5537184l37847u14
47. Torres-Lima P, Canabal-Cristiani B, Burela-Rueda G (1994) Urban sustainable agriculture: The paradox of the Chinampa system in Mexico City. Agric Hum Values 11(1):37–46. https://doi.org/10.1007/BF01534447
48. Akter S (2024) Harnessing technology for environmental sustainability: Utilizing AI to tackle global ecological challenges. J Artif 2(1):49–57. https://doi.org/10.60087/jaigs.v2i1.p57
49. Ramirez Lopez LJ, Grijalba Castro AI (2020) Sustainability and resilience in smart city planning: A review. Sustainability 13(1):181. https://doi.org/10.3390/su13010181
50. Mori K, Christodoulou A (2012) Review of sustainability indices and indicators: Towards a new city sustainability index (CSI). Environ Impact Assess Rev 32(1):94–106. https://doi.org/10.1016/j.eiar.2011.06.001
51. Yaashikaa PR, Kumar PS (2022) Bioremediation of hazardous pollutants from agricultural soils: A sustainable approach for waste management towards urban sustainability. Environ Pollut 312:120031. https://doi.org/10.1016/j.envpol.2022.120031
52. Tiwari AK, Mishra H, Nishad DC (2024) Market dynamics and consumer perceptions of organic produce in contemporary agriculture. Adv Agric Sci 24(4):120
53. Sintusingha S (2011) Bangkok's urban evolution: Challenges and opportunities for urban sustainability. Megacities, pp 133–161. https://doi.org/10.1007/978-4-431-99267-7_7
54. Singh R, Singh H, Raghubanshi AS (2019) Challenges and opportunities for agricultural sustainability in changing climate scenarios: A perspective on Indian agriculture. Trop Ecol 60(2):167–185. https://doi.org/10.1007/s42965-019-00029-w
55. Dewan AM, Kabir MH, Nahar K, Rahman MZ (2012) Urbanisation and environmental degradation in Dhaka metropolitan area of Bangladesh. Int J Environ Sustain Dev 11(2):118–147. https://doi.org/10.1504/IJESD.2012.049178
56. Gómez-Baggethun E, Barton DN (2013) Classifying and valuing ecosystem services for urban planning. Ecol Econ 86:235–245. https://doi.org/10.1016/j.ecolecon.2012.08.019
57. Islam R, Siwar C (2012) The analysis of urban agriculture development in Malaysia. Adv Environ Biol 6(3):1068–1078
58. Tapia C, Randall L, Wang S, Aguiar Borges LA (2021) Monitoring the contribution of urban agriculture to urban sustainability: An indicator-based framework. Sustain Cities Soc 74:103130. https://doi.org/10.1016/j.scs.2021.103130
59. Seabrook L, Mcalpine CA, Bowen ME (2011) Restore, repair or reinvent: Options for sustainable landscapes in a changing climate. Landsc Urban Plan 100(4):407–410. https://doi.org/10.1016/j.landurbplan.2011.02.015

60. Tiwari AK, Mishra H (2024) Sustainable agriculture: The future of world. In: Patel DK, Dwivedi S, Dubey A, Kumar A, Ali W (Eds) Agricultural Transformation in the Modern Era. Elite Publishing House, pp 167–192.
61. Fu BJ, Zhuang XL, Jiang GB, Shi JB, Lu Y (2007) Feature: Environmental problems and challenges in China. Environ Sci Technol 41(22):7597–7602. https://doi.org/10.1021/es0726431
62. Schellenberg T, Subramanian V, Ganeshan G, Tompkins D, Pradeep R (2020) Wastewater discharge standards in the evolving context of urban sustainability-the case of India. Front Environ Sci 8:30. https://doi.org/10.3389/fenvs.2020.00030
63. Bellezoni RA, Meng F, He P, Seto KC (2021) Understanding and conceptualizing how urban green and blue infrastructure affects the food, water, and energy nexus: A synthesis of the literature. J Cleaner Prod 289:125825. https://doi.org/10.1016/j.jclepro.2021.125825
64. Duží B, Frantál B, Simon Rojo MS (2017) The geography of urban agriculture: New trends and challenges. Moravian Geogr. Rep. 25(3):130–138. https://doi.org/10.1515/mgr-2017-0012
65. Garnett T (2013) Food Sustainability: Problems, Perspectives and Solutions. Proc Nutr Soc 72(1):29–39. https://doi.org/10.1017/S0029665112002947
66. Mokthsim N, Salleh KO (2014) Malaysia's efforts toward achieving a sustainable development: Issues, challenges and prospects. Procedia Soc Behav Sci 120:299–307. https://doi.org/10.1016/j.sbspro.2014.02.107
67. Singh M, Mishra H (2023) Medicinal plant farming to improve farmers' economy. Agriallis 30–36
68. Maye D (2019) "Smart food city": Conceptual relations between smart city planning, urban food systems and innovation theory. City Cult Soc 16:18–24. https://doi.org/10.1016/j.ccs.2017.12.001
69. Wiskerke JS, Viljoen A (2012) Sustainable urban food provisioning: Challenges for scientists, policymakers, planners and designers. In: Sustainable Food Planning: Evolving Theory and Practice, pp 19–36. Wageningen Academic
70. Dimnwobi SK, Okere KI, Onuoha FC, Ekesiobi C (2023) Energy poverty, environmental degradation and agricultural productivity in Sub-Saharan Africa. Int J Sustain Dev World Ecol 30(4):428–444. https://doi.org/10.1080/13504509.2022.2158957
71. Allen A, da Silva N, Corubolo E (1999) Environmental problems and opportunities of the Periurban interface and their impact upon the poor
72. Quon S (1999) Planning for urban agriculture: A review of tools and strategies for urban planners. Cities Feeding People Series; rept; vol. 28, 1999
73. Carley M, Christie I (2017) Managing sustainable development, Routledge
74. Olanipekun IO, Olasehinde-Williams GO, Alao RO (2019) Agriculture and environmental degradation in Africa: The role of income. Sci Total Environ 692:60–67. https://doi.org/10.1016/j.scitotenv.2019.07.129
75. Nishad DC, Mishra H, Tiwari AK, Mishra D (2024) Post-harvest Management: Enhancing food security and sustainability. Adv Agric Sci 24(4):136
76. Mutisya E, Yarime M (2014) Moving towards urban sustainability in Kenya: A framework for integration of environmental, economic social and governance dimensions. Sustain Sci 9(2):205–215. https://doi.org/10.1007/s11625-013-0223-7
77. Cilliers S, Cilliers J, Lubbe R, Siebert S (2013) Ecosystem services of urban green spaces in African countries—Perspectives and challenges. Urban Ecosyst 16(4):681–702. https://doi.org/10.1007/s11252-012-0254-3
78. Sekovski I, Newton A, Dennison WC (2012) Megacities in the coastal zone: Using a driver-pressure-state-impact-response framework to address complex environmental problems. Estuarine Coast Shelf Sci 96:48–59. https://doi.org/10.1016/j.ecss.2011.07.011
79. Balaban O (2012) The negative effects of construction boom on urban planning and environment in turkey: Unraveling the role of the public sector. Habitat Int 36(1):26–35. https://doi.org/10.1016/j.habitatint.2011.05.003
80. Haq SMA (2011) Urban green spaces and an integrative approach to sustainable environment. J Environ Prot 02(5):601–608. https://doi.org/10.4236/jep.2011.25069

81. Mishra H, Tiwari AK, Nishad DC (2024) Economic viability of sustainable agriculture practices in modern farming. Adv Agric Sci 24(4):105
82. Haaland C, van Den Bosch CK (2015) Challenges and strategies for urban green-space planning in cities undergoing densification: A review. Urban Urban Greening 14(4):760–771. https://doi.org/10.1016/j.ufug.2015.07.009
83. Gomiero T (2016) Soil degradation, land scarcity and food security: Reviewing a complex challenge. Sustainability 8(3):281. https://doi.org/10.3390/su8030281
84. Koger SM (2011) The Psychology of Environmental Problems: Psychology for Sustainability. Psychology Press
85. Godson-Ibeji CC, Chikaire JU (2016) Consequences of environmental pollution on agricultural productivity in developing countries: A case of Nigeria. Int J Agric Food Res 5(3)
86. Metu AG, Okeyika KO, Maduka OD (2016) Achieving sustainable food security in Nigeria: Challenges and way forward. In: 3rd International Conference on African Development Issues (CU-ICADI), pp 182–187
87. Rozelle S, Huang J, Zhang L (1997) Poverty, population and environmental degradation in China. Food Policy 22(3):229–251. https://doi.org/10.1016/s0306-9192(97)00011-0
88. Benke K, Tomkins B (2017) Future food-production systems: Vertical farming and controlled-environment agriculture. Sustain Sci Pract Policy 13(1):13–26. https://doi.org/10.1080/15487733.2017.1394054
89. Gar-on Yeh AGO, Wu F (1999) The transformation of the urban planning system in China from a centrally planned to transitional economy. Prog Plann 51(3):167–252. https://doi.org/10.1016/S0305-9006(98)00029-4
90. Mishra H, Mishra D (2024) Economic evaluation of UAV-based soil sampling approaches. In: Applications of Computer Vision and Drone Technology in Agriculture 4.0, Springer Nature Singapore, pp 271–291
91. Sutthichaimethee P, Sawangdee Y (2016) Indicator of environmental problems of agricultural sectors under the environmental modeling. J Ecol Eng 17(2):12–18. https://doi.org/10.12911/22998993/62280
92. Steinfeld H, Gerber P, Wassenaar TD, Castel V, De Haan C (2006) Livestock's long shadow: environmental issues and options. Food & Agriculture Org
93. Tyagi S, Garg N, Paudel R (2014) Environmental degradation: Causes and consequences. Eur Res 81(8–2):1491
94. Bouma J, Van Altvorst AC, Eweg R, Smeets PJAM, Van Latesteijn HC (2011) The role of knowledge when studying innovation and the associated wicked sustainability problems in agriculture. Adv Agron 113:293–323. https://doi.org/10.1016/B978-0-12-386473-4.00006-3
95. Midler E (2022) Environmental Degradation: Impacts on Agricultural Production. IEEP-Policy Brief. https://ieep.eu/publications/environmental-degradation-impacts-onagricultural-production
96. Hasnat GT, Kabir MA, Hossain MA (2018) Major environmental issues and problems of South Asia, particularly Bangladesh. Handb Environ Mater Manag 1–40
97. Merson J, Attwater R, Ampt P, Wildman H, Chapple R (2010) The challenges to urban agriculture in the sydney basin and lower blue mountains region of Australia. Int J Agric Sustain 8(1–2):72–85. https://doi.org/10.3763/ijas.2009.0464
98. Carter FW, Turnock D (Eds) Environmental problems of east central Europe
99. Menyuka NN, Sibanda M, Bob U (2020) Perceptions of the challenges and opportunities of utilising organic waste through urban agriculture in the Durban South Basin. Int J Environ Res Public Health 17(4):1158. https://doi.org/10.3390/ijerph17041158
100. Mishra H, Mishra D (2023) Artificial intelligence and machine learning in agriculture: Transforming farming systems. Res Trends Agric Sci 1:1–16
101. Pribadi DO, Pauleit S (2016) Periurban agriculture in Jabodetabek metropolitan area and its relationship with the urban socioeconomic system. Land Use Policy 55:265–274. https://doi.org/10.1016/j.landusepol.2016.04.008

102. Musvoto C, Nortje K, De Wet B, Mahumani BK, Nahman A (2015) Imperatives for an agricultural green economy in South Africa. S Afr J Sci 111(1/2):1–8. https://doi.org/10.17159/sajs.2015/20140026
103. Duží B, Tóth A, Bihuňová M, Stojanov R (2014) Challenges of urban agriculture: Highlights on the Czech and Slovak Republic specifics. Curr Chall Cent Eur Soc Environ 82–107
104. Giles-Corti B, Vernez-Moudon A, Reis R, Turrell G, Dannenberg AL, Badland H, Foster S, Lowe M, Sallis JF, Stevenson M, Owen N (2016) City planning and population health: A global challenge. Lancet 388(10062):2912–2924. https://doi.org/10.1016/S0140-6736(16)30066-6
105. Sousa R, Batista D (2013) Urban agriculture: The allotment gardens as structures of urban sustainability. InTech
106. Chowdhury T, Furedy C (1994) Urban sustainability in the third world: A review of the literature; Institute of Urban Studies
107. Zhang X (2016) Sustainable urbanization: A Bi-dimensional Matrix Model. J Cleaner Prod 134:425–433. https://doi.org/10.1016/j.jclepro.2015.08.036

Wastewater Management and Urban Water Sustainability

Manglam Soni, Sonal Singh, Shashank Gupta, Kuldip Dwivedi, and Nidhi Shukla

Abstract The rapid expansion of urban areas has intensified the challenges associated with water resource management, particularly its role in the context of wastewater management, which is vital for achieving urban water sustainability. This paper examines the diverse sources of wastewater, including domestic, agricultural, and industrial contributions, and categorizes them into stormwater runoff, sewage, and non-sewage wastewater. It underscores the critical need for the advanced waste water treatment systems to address the complex composition of effluents, which often contain harmful pollutants like as heavy metals, organic and inorganic compounds, and pathogens. The study highlights various sustainable practices and innovative technologies that can enhance wastewater management, including low-impact development strategies like permeable pavements and green roofs, as well as water reclamation and reuse techniques. The role of microorganisms in bioremediation processes is also discussed, emphasizing their potential in the biosorption and bioaccumulation of toxic substances from wastewater. Furthermore, the paper advocates for a holistic approach that integrates community engagement, effective policy frameworks, and investment in infrastructure to foster sustainable water management practices. By promoting water conservation, nutrient recovery, and energy management, urban areas can mitigate the adverse effects of wastewater on freshwater resources and ecosystems. Ultimately, this research objective is to deliver a complete understanding of the importance of wastewater management in urban settings, offering insights into sustainable applications that in spite of continuous environmental problems, it can guarantee the robustness and well-being of metropolitan water systems.

Keywords Wastewater · Urbanization · Industrial · Heavy metals · Sustainbiity

---

M. Soni
Amity Institute of Biotechnology, Amity University Madhya Pradesh, Gwalior, India

S. Singh · K. Dwivedi · N. Shukla (✉)
Department of Environmental Science, Amity School of Life Science, Amity Univeristy Madhya Pradesh, Gwalior 474001, India
e-mail: nidthesaviour@gmail.com

S. Gupta
Department of Civil Engineering, ITM University, Gwalior 474005, India

J. A. Parray et al. (eds.), *Sustainable Urban Environment and Waste Management*, Urban Sustainability, https://doi.org/10.1007/978-981-96-1140-9_4

1 Introduction

If the demand of water for the public is not mitigated by natural sources or its absence in natural environment, the last option is the treatment of water using various techniques without effecting other properties of water related to their physical and chemical phenomena, thus it could be beneficial for all living things. Urbanization has brought tremendous challenges to water resources management. One of the most pressing concerns is the management of wastewater in a way that promotes urban water sustainability. The rapid expansion of urban areas, coupled with the rising demand for clean water, has necessitated the improvement of advanced wastewater management systems. These systems are crucial to reducing pollution, conserving water, and fostering the overall health of urban ecosystems. Wastewater get daily discharges from various sources like household activities, agricultural and industrial activities with heavy load of hydraulic and diverse compositions [1, 2]. The effluents contain various organic, inorganic, carbon, nitrogenous organics, dissolved, suspended solids and heavy metals [3, 4]. Wastewater treatment is the technology and process which helps to treat remove readily impurities or pollutants that are existing in wastewater to guarantee and public hygiene. Thus, water waste Managing therefore means to handling with wastewater to defend the environment and confirms public health safety, economic, social and political [5].

Wastewater management plays a significant role in achieving urban water sustainability. Effective management ensures that water is treated, reused, or safely discharged into the environment, thus reducing the impact on freshwater sources. This chapter explores the importance of wastewater management, the challenges faced by urban areas, and sustainable practices to ensure a balance between urban growth and water sustainability.

2 Type of Wastewater

In day-to-day life we divided into two components like sewage wastewater and non-sewage wastewater [6]. Sewage wastewater consists of release through the domestic and homemade events like wastewater from different sources such as house, education centre, hospitals, institutes, public washrooms, etc. which includes urine and stools or faeces comes under the category of sewage wastewater. The rest wastewater generated from like industries, factories, etc. have come under the category of non-sewage wastewater. At the time of rain or storm all the water that overflow as flood water is also considered as non-sewage wastewater. All the day-to-day activities done by humans are also water related like bathing, cooking and cleaning are related to water which create high amount of water waste. This shows the importance of wastewater management and treatment.

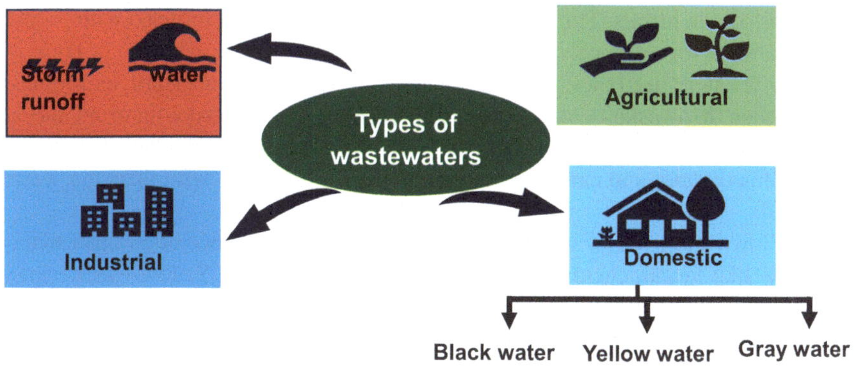

Fig. 1 Categories of wastewaters generated by various Domestic and Non-domestic wastewater

As discussed earlier, the waste water is combination of several different sources. Thus, the handling process of waste water and the specific treatment for that individual source is needed. The treatment process is deeply divided into many types and subtypes as per the different sources. The four major types of wastewaters are stormwater runoff, domestic, agricultural and industrial [7] (Fig. 1).

(a) Storm water Runoff wastewater: Stormwater Runoff wastewater includes heavily rainwater, water of storm, water from flood flows, etc. Mainly it includes the water that does not infiltrate into ground due to hard strata and the water flows onto the surface of the roads or on open areas. The storm water can be considered as the biggest cause of water pollution as the pollutants that it carries has a high variety, like plastics, oils, chemicals, pesticides, herbicides, heavy metals and even various types of diseases, wash into stormwater runoff from roadways, businesses building locations and other places. Untreated stormwater runoff typically empties into neighbouring natural waterways including ponds, rivers, streams, and lakes either directly or via channelled drains. Every life depends either directly or indirectly on natural waterways for their survival, therefore this contaminated water threatens not just aquatic life but the entire ecosystem.

(b) Domestic wastewater: The wastewater generated by households derived from human activities are called as domestic wastewater. The domestic wastewater divided into two categories:

(1) Toilet waste: The liquid generated from laundry, bathing, washing, etc.
(2) Wastewater generated through household activities like cooking, etc. [8].

Based on the source of wastewater are categorised into three sub types.

(i) Grey water: Greywater contains less contaminated water than other types of wastewaters, which contains washing machine, baths and bathrooms [9]. In simply grey wastewater consist of bathing liquids, soap, detergent and different types of chemicals. The grey water is not pathogenic in nature and with proper treatment for chemicals it can be re-use it again.

(ii) Blackwater: the black water consisting of various effluents of bathrooms, kitchen dishwashes, etc. the black wastewater contaminated with urine, faecal matter, discarded food pieces and different types of chemicals. It is most polluted water and can cause disease if it gets mixed up with drinking water [10].

(iii) Yellow water: yellow water contains only urine and absence of all other materials like faecal, chemicals, grey water and black water. Yellow water marks the absent of any food particles, etc. [11].

(iv) Agricultural wastewater: In many watersheds, agricultural runoff is thought to be a significant cause of water contamination. When too much water goes off the fields during surface irrigation, it's common to refer to agricultural wastewater as irrigation tailwater [12]. The running of extra water over the field causes the eradication of minerals and nutrients from the fertile soil. In addition to those agrochemicals such as fertilizers, pesticides, herbicides, crop residuals, animal waste, etc. are the sources of the agricultural wastewater [13]. Many practices are undertaken to avoid prevent the increment of the agricultural wastewater through other contaminants [14, 15].

(v) Industrial wastewater: Industrial wastewater consisting dissolved and suspended discharged from the manufacturing plant, testing labs, chemical room, other commercial activities, etc. [16]. On the nature of the industry, wastewater contains different types of the contaminants (Table 1).

3 Some Highlighted Study on the Wastewater Treatment and Water Restorage

(a) Kumar and Arora [33] done discussion on the application of biotechnology in which effective ligninolytic enzymes, manganese peroxides (MnPs) that are important for the preliminary breaking a broad variety of recalcitrant contaminants which includes lignin, toxic compounds, dyes, mycotoxins, antibiotics, hydrocarbon discharge in the wastewater. However, MnPs uses in the paper pulp industry to decrease the amount of lignin factors and chemicals in biopulping and biobleaching in industries.

(b) Efrain Merma [34] worked on the recognition, occurrence and ecotoxicity in aquatic and terrestrial ecosystem of medicines tested to treat the COVID- 19 disease. Corresponding to the various data gathered by them shows tested drugs are existing in different environmental spaces like domestic water, common effluents treatment plants, hospital waste, etc. all mixed up in water and get contaminated with various types of bacteria, fungus, etc. which can cause severe disease if gets mixed up with drinking water. Thus, to cure the wastewater with algae action, photocatalysis, photodegradation, photo-Fenton and advanced

Table 1 Industries with their effluents

S. No	Industry	Waste commonly found	References
01	Paper Pulp Industry	Suspended Matters, Lignin's, Tannins, Colours, Sterols, etc	[17, 18]
02	Paint Industries	Lead, Chromium, Zinc, (VOCs) Volatile Organic Compounds	[19], 20, 21
03	Textile	Urea, Iron, Chromium, Hydrogen Peroxide, etc	[22–24]
04	Metal Working	Perfluoro Octane Sulfonates (PFOS), Ammonium, Nitrogen, Cyanide, Phenol, Oil and Grease	[25, 26]
05	Dye Industries	Copper, Colours, Sulphides, Formaldehydes	[27, 28]
06	Pharmaceutical bases Industries	Cadmium, Nickel, Synthetic surfactants	[29, 27]
07	Petrochemical	Long chain hydrocarbon, Phenolic derived compounds, Nitrobenzene, Alkanes, Chloroalkanes, High Salt, etc	[29]
08	Plastic	Perfluorooctanoic Acid (PFOA), Phenols, Vinyl chloride, Cadmium, styrene	[26, 30, 31]
09	Winery	Variable Flows and Organic Matter, Nitrogen and Suspended Solid	[32]
10	Battery producer	Heavy Metals, Oils, Lubricates, etc	[32]

oxidation process seems to best method for the treatment of the contaminated wastewater rather than treating with chemicals or antibiotics.

(c) Prakash et al., [35] discovered the incorporating the usage of macrophytes (*Typha* sp. and *Phragmites* sp.) and highly indicative zooplanktons (*Monia macrocopa* and *Brachinous angularis*) are constructed in marshland for the cleansing of wastewater after 3 years study, there is increased up yield in the approx. The observation of decrease by 92% was seen in biochemical oxygen demand and decrease of 56% was seen in total bacterial count, while the content of dissolved oxygen was observed to be increased significantly.

(d) Bera et al., [36] established a highly potent bacterial consortium, SB92 consist of *K. rosea, B. flexus, B. tequilensis* and *P. stutzeri* which helps to decoloruri-sation and breakdown of orange acid at the rate of 30 mg/ml within the slight nourishment in static conditions at 23 h in laboratory. More so ever SPB92 helps to decolorised 05 azo dye which possess similar structure to the acid orange. This signify the importance of SPB92 in management of textile wastewater discharged in open environment without adequate supply of the nutrients.

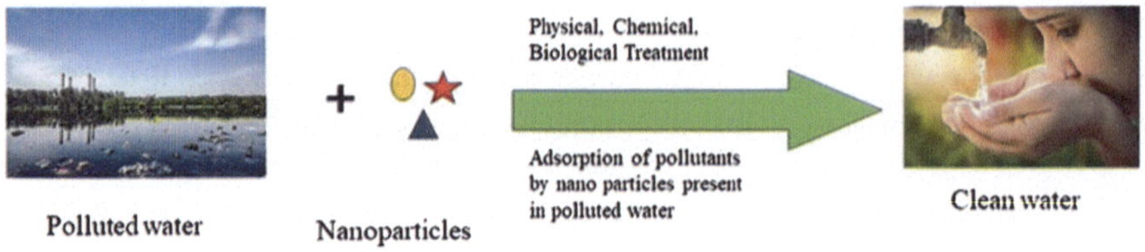

Fig. 2 Use of nanoparticle's in the cleaning of wastewater

(e) Deepa et al., [37] developed ecofriendly and sustainable method to treat the E. coli bacteria using silver nanoparticles made from the unutilised vegetables waste from peels of Lagenaria and Pisum sativum possess antimicrobial activity thus help to control the growth of *E. coli* and treat wastewater. Silver nanoparticles diameter around 30–55 mm are made from the sustainable approach thus they are not burden on the environment like other nano particles possess.

(f) Singh et al., [38] mentioned that various metallic nano particles, including Ag, Au, Zn, Fe, Ti TiO_2, and Sio_2 was synthesised with diverse conformations like shape and sizes, by using a variability of plant products like leaves, roots, bark, and flowers. If we compare this synthesis with different methods like physical and chemical methods then it is observed to be very economical as it has lower price and along with this it is also friendly towards environment, and the process is also more sustainable. During its synthesizing, the role of capping and reducing agents is performed by special enzymes and proteins that are found in plant extract. Nano particles generated from plants can be used to reduce the amount of pollutants present in waste water whether organic or inorganic. High range of contaminant such as dyes, heavy metals, can be treated with these plant based nano particles (Fig. 2).

(g) Mallick et al., [39] stated that sustainable water resource management in Kurseong's hilly region, emphasizing watershed management to address groundwater scarcity. Local inhabitants, reliant on rainfed agriculture and natural jhoras for water, face risks due to climate variability and disrupted water cycles. Recent changes have led to Jhora's becoming seasonal or extinct, threatening water availability for drinking, household use, and irrigation. Rainfall data from 13 locations show adverse impacts on productivity due to groundwater shortages. The study advocates for an integration of watershed management approach to conserve groundwater and mitigate water scarcity for local communities and agriculture.

4 Conventional Wastewater Treatment

Traditional wastewater treatment consisting of a fusion of different types of physical, chemical and biological methods for the treatment of wastewater. It helps to remove solids, organic solids and help to remove nutrients from the wastewater [40].

(a) Preliminary Treatment: The primary aim for preliminary treatment is the exclusion of the solid coarse and extra huge material readily present in the raw wastewater. After the completion of Preliminary treatment, primary treatment is done in which the removal of suspended and floating solids are removed. Preliminary treatment helps to remove bits of wood, cloths, plastic, compost, etc. along with faecal matter. In preliminary treatment includes also removal of heavy inorganic solid like glass, metals, etc.
(b) Primary Treatment: Primary treatment helps to eliminate the biological or carbon based and inorganic matter by the physical process methods of deposit and floatation. It helps to remove 25–50% of the biochemical oxygen demand (BOD) [52], approximate 55–75% of the Total suspended solids (SS) and 60% of the oil and grease. The Removal of heavy metals, organic phosphorus and organic nitrogen can also be done with the help of primary treatment. The effluents which are generated from primary treatment are called as primary effluents. Primary treatments include clarifiers or settling tanks which helps to remove of inorganic solids that set when left for some time from the wastewater. After the primary treatment the outlet of the treated water still contains colloidal solids and organic and inorganic in the form of dissolved solids. To maintain the water quality and standard water needs to treat further. Sometimes organic matter in waste water is not removed by the primary treatment alone, but additionally it requires secondary treatment to remove other components.
(c) Secondary Treatment: The main objective of the secondary treatment is the extend treatment of the effluents that have gone through the process of primary treatment. The water contains suspended solids and residual of organic matter. The effluent that come for the secondary treatment is mainly composed of 70% of dissolved solids but about 25% of suspended solids and 5% of colloidal solids are also present that are not purely removed by the primary treatment. Secondary treatment process involves the biological treatment by using various microorganisms in a controlled condition of environment. In secondary treatment, the bio-organisms are used for the aerobic process in which oxygen is supplied to the microorganisms so that they break down the organic debris into solids and then the removal process becomes easier.

5 Toolkit to Achieve Increase Urban Water and Wastewater Management Sustainability

(a) Stormwater management and rainwater harvesting: Stormwater management mainly concerns with the slowdown of flow of surface water. There are many techniques that can be done to stop the rain water at various spaces or to slow down the speed of flow so that the natural treatment could occur. At slow pace it will be easier to remove pollutants. If the storm water is captured at a place, then it gets more time for the infiltration process and the evaporation and evapotranspiration process can also be increased so that the water will return to the environment. The stormwater can be achieved by permeable pavements, green roofs, rain gardens, bioretention basins are also known as low impact development [41].

(b) Water conservation: the application of technologies that provide expected serve with reduced water use, potentially coupled with behaviour changes. Low flow shower heads, toilets and washing machines drip irritation are the best example of the water conservations [42].

(c) Water reclamation and reuse: The waste water that has been treated is water reclaimed and to use it again is the reuse of water. Water reclaimation and reuse depend upon the quantity of water that have been retained as well as the quality of water.

(d) Energy management: The flow of waste water can be used for the generation of energy. The amount of organic matter and nitrogen can be converted into chemical energy and then into electrical energy or mechanical or even thermal energy. Treatment with anaerobic treatment, thermal combustion with energy recovery and can be use in microbial fuel.

(e) Nutrients recovery: Waste water not just contains pollutants but also have some matter that can be used for better purpose. If the nutrients can be extracted from waste water, then it can be converted as fertilizers and then used for agricultural processes. It can be use in land application of biosolids.

(f) Source separation: If the waste can be separated as per their source, then different characteristics can be determined and their potential uses can be achieved.

6 Role of Microorganism of in the Treatment of Wastewater Management in Urban

Wastewater can be treated with broadly with two aspects i.e., biosorption and bioaccumulations. Both these methods use in the bioremediation of wastewater. Biosorption comprises of instantly adsorption of substances without expenditure of energy, it is reversible process in nature. Biosorption can be on a biomass of living as well as non-living. Therefore, this method is economical and cheap method. The biomass can be used multiple times for the treatment of the wastewater. While bioaccumulation in which the gathering of toxic compounds takes places by the continuous process from the environment. Bioaccumulation can only be performed by the living biomass only and thus makes this process costly and cannot be reused repetitively because of the life cycles of the living organism [43–45]. Bacteria are the potent microorganisms which helps in the process of bioremediation through biosorption and bioaccumulation.

Various selected species of bacteria *like Endophytes, Arthrobacter, Pseudomonas* sp., *Bacillus sp. D vulgaris, S. marcescens,* etc. have the tendency to decrease the toxicity level of Cr (VI) into less toxic Cr (III) form, *A. psychrolactophilus* Sp 114 helps to reduce the concentration of proteins which are majorly used by industry in the sewage treatment. According to various reports Cyanobacteria helps to remove heavy metals, colours particles and crude oil by biosorption process. Nitrates can easily be removed by cyanobacteria. Common occurring strain of cyanobacteria like *Synechococcus* sp. and *Synechocystis minima* helps to remove various contaminants from water. For the treatment of domestic waste water and industrial waste water A. variabilis and Tolypothrix ceytonia are also widely used along with cyanobacteria Anabaena oryzae. Phosphorus and nitrogen can be removed by the *P. tenue* and *P. bohneri. A. variabilis* helps to remove heavy organic load while solid particles can be removed by *T. ceytonica* and *A. variabilis. A. subcylindrica* and *N. muscorum* helps in the accumulation of several heavy metal that include cobalt, manganese, and copper, from sewage wastewater. For the breakdown of complex crude oil some Species like *Aphanocapsu sp, Terenbans* and *Oscillatoria salina* are being used. *Anabaena sp.* and *Nostoc ellipsosporum* helps to degrade pesticides effectively. Dyes from textile industries can effectively removed by *O. formosa.* and Phosphorus can be removed by *R. sphaeroides* [46, 47].

Algae are considered as natural purifier for the purification water. They help in absorption of poisonous and radioactive metal ions from wastewater. Algae have the tendency to rapidly grow which results in absorption of more amount of nutrients from wastewater, thus this property of algae can be helpful in remediation of various nutrients. Therefore, the treatment of sewage wastewater can be done with the help of this technology which can act as an alternate of secondary and tertiary treatment and will also be economical. Algae are also helpful in removal of dye and contaminants from textile wastewater and supports the growth of algae because of rich in nutrients.

C. vulgaris helps in reduction of the effluents responsible for increment of biological oxygen demand and chemical oxygen demand of the waste water. It also helps in control the production of citric acid and diluted forms of alcohol from water waste. *C. vulgaris* and *S. quadricauda* both are found efficient in the exclusion of nitrate. *C. vulgaris* also helpful in the removal of phosphorus through remediation. Removal of heavy metal and ions from wastewater through algae biobased biosorption are considered as eco-friendly. Thus, this methos considered as economic, environment friendly and sustainable methods for the treatment of wastewater. Live algae like *C. Lentillifera* and *C. scalpelliformis* were effectively found to remove basic dyes through biosorption. Algae like *Phormidiumis* known to convert textile toxic dyes to their nontoxic form by bioconversion [48, 47].

Filamentous fungi can be used in waste water treatment as it uses several heavy metals like Nickel, Zinc, Cadmium, Lead, Thorium, Radium and Uranium for their growth and many other metabolically activities. These Fungal biomasses can also be used for the cleaning of industrial effluent wastewater as it helps in removing the radionuclides that emit radiations. It has been observed that the listed heavy metals can be absorbed by various species of fungi like *Rhizopus, Mucor, Penicillium, Aspergillus, Saccharomyces,* and *Fusarium.* A variety of fungi named "White rot fungi" are found to be very effective in the decrease of phenolic compounds present in wastewater of industrial. *C. versicolor, F. trogii, P. chrysosporium* and *P. pulmonarius* can decolorize the molasses and are able to decrease the amount of chemical oxygen demand (COD). *F. ventricosum, P. chrysosporium, A. terreus, C. oxysporum, M.*

thermohyalospora, and *T. harzianum* were founded effectively to degrade endosulfan molecules. Fungi belongs to class *zygomycetes* and species of *P. chrysogenum, S. apiospermum, P. digitatum* and *F. solani* can degrade polychlorinated biphenyls (PCB).

Pinus with S. bovines and R. are combined and that solution is used for the removal of cadmium from waste water. The list of fungi that is used for the treatment of waste water is not just limited to these listed fungi but some that are plant associated fungi like *T. Hirsuta, T. viride, A. nidulans, B. adusta, F. trogii, I. lacteus, P. ostreatus,* have also been used popularly. In the textile industry the wastewater consist of different colours is treated and decolorization can be done with the help of these fungi. Yeast like *T. cutaneum, C. tropicalis* and *Saccharomyces sp.* are majorly used in the bioremediation of oil in wastewater. They are known to decrease the COD levels and along with the efficient removal of mono and polyphenols compounds. They can also convert toxic chromophore compounds produced into non-toxic forms in waste water generated from industry. 83, 85, 89 [49–51].

7 Conclusion

The conclusion on "Wastewater Management and Urban Water Sustainability" emphasizes the importance of integrating innovative wastewater treatment technologies and sustainable practices to address the growing challenges of urban water scarcity and pollution. It highlights that effective wastewater management not only helps in reclaiming water for reuse but also plays a crucial role in protecting the environment and public health.

Key points from the conclusion include:

1. **Sustainable Practices**: The need for adopting sustainable water management practices that include water conservation, reclamation, and reuse to ensure a reliable water supply for urban populations.
2. **Technological Innovations**: The role of advanced technologies, such as bioremediation and the use of nanoparticles, in enhancing the efficiency of wastewater treatment processes and reducing environmental impacts.
3. **Community Involvement**: The importance of engaging local communities in water management strategies to foster a sense of ownership and responsibility towards water resources.
4. **Policy and Governance**: The necessity for supportive policies and governance frameworks that promote sustainable water management practices and encourage investment in wastewater treatment infrastructure.

Overall, the study advocates for a holistic approach to wastewater management that combines technological advancements, community engagement, and effective policymaking to achieve urban water sustainability and resilience against climate change impacts.

References

1. Tom AP, Jayakumar JS, Biju M, Somarajan J, Ibrahim MA (2021) Aquaculture wastewater treatment technologies and their sustainability: a review. Energy Nexus 4:100022
2. Lata S (2021) Sustainable and eco-friendly approach for controlling industrial wastewater quality imparting succour in water-energy nexus system. Energy Nexus 3:100020
3. Wang J, Ji Y, Zhang F, Wang D, He X, Wang C (2019) Treatment of coking wastewater using oxic-anoxic-oxic process followed by coagulation and ozonation. Carbon Resour Convers 2:151–156
4. Senfter T, Fritsch L, Berger M, Kofler T, Mayerl C, Pillei M, Kraxner M (2021) Sludge thickening in a wastewater treatment plant using a modified hydrocyclone. Carbon Resour Convers 4:132–141
5. Metcalf and Eddy, Inc (1991) "Wastewater Engineering": Treatment Disposal and Reuse, third edition. New York: McGraw-Hill
6. Schoen ME, Ashbolt NJ (2010) Assessing pathogen risk to swimmers at non-sewage impacted recreational beaches. Environ Sci Technol 44(7):2286–2291. https://doi.org/10.1021/es903523q
7. Amoatey P, Bani R (2011) Wastewater management. In waste water—Evaluation and management. InTech. https://doi.org/10.5772/16158
8. Mara D (2013) Domestic wastewater treatment in developing countries. Routledge. https://doi.org/10.4324/9781849771023
9. Al-Jayyousi OR (2003) Greywater reuse: Towards sustainable water management. Desalination 156(1–3):181–192. https://doi.org/10.1016/s0011-9164(03)00340-0
10. Yin H, Xie M, Zhang L, Huang J, Xu Z, Li H, Jiang R, Wang R, Zeng X (2019) Identification of sewage markers to indicate sources of contamination: Low cost options for misconnected non-stormwater source tracking in stormwater systems. Sci Total Environ 648:125–134. https://doi.org/10.1016/j.scitotenv.2018.07.448
11. Larsen TA, Udert KM, Lienert J (2013). Source separation and decentralization for wastewater management. https://doi.org/10.2166/9781780401072
12. Pedrero F, Kalavrouziotis I, Alarcón JJ, Koukoulakis P, Asano T (2010) Use of treated municipal wastewater in irrigated agriculture—Review of some practices in Spain and Greece. Agric Water Manag 97(9):1233–1241. https://doi.org/10.1016/j.agwat.2010.03.003
13. Evans AE, Mateo-Sagasta J, Qadir M, Boelee E, Ippolito A (2019) Agricultural water pollution: Key knowledge gaps and research needs. Curr Opin Environ Sustain 36:20–27. https://doi.org/10.1016/j.cosust.2018.10.003
14. Ongley ED (1996) Control of water pollution from agriculture (Vol. 55). Food & Agriculture Org
15. Vymazal J (2009) The use constructed wetlands with horizontal sub-surface flow for various types of wastewater. Ecol Eng 35(1):1–17. https://doi.org/10.1016/j.ecoleng.2008.08.016
16. Munter R (2003) Industrial wastewater characteristics. The Baltic University Programme (BUP), Sweden, p 185
17. Bui HN, Chen YC, Pham AT, Ng SL, Lin KA, Nguyen NQV, Bui HM (2022) Life cycle assessment of paper mill wastewater: a case study in Viet Nam. Water Sci Technol: J Int Assoc Water Pollut Res 85(5):1522–1537. https://doi.org/10.2166/wst.2022.049
18. Rajni S, Satish K, Chhaya S (2011) Tertiary treatment option for pulp and paper mill wastewater to achieve effluent recycling. IPPTA: Q J Indian Pulp Pap Tech Assoc 23:155–159
19. Aniyikaiye TE, Oluseyi T, Odiyo JO, Edokpayi JN (2019) Physico-Chemical analysis of wastewater discharge from selected paint industries in Lagos, Nigeria. Int J Environ Res Public Health 16(7):1235. https://doi.org/10.3390/ijerph16071235
20. Mostafa MK, Peters RW (2016) Reuse paint wastewater in the manufacture of cement bricks and tiles. J Mater Cycles Waste Manage 19(2):840–850. https://doi.org/10.1007/s10163-016-0485-0
21. Shazly MAE, Hasanin EA, Kamel MM (2010) Appropriate technology for industrial wastewater treatment of paint industry. Am Eurasian J Agric Environ Sci 8(597):601

22. Ćurić I, Dolar D, Bošnjak J (2021) Reuse of textile wastewater for dyeing cotton knitted fabric with hybrid treatment: Coagulation/sand filtration/UF/NF-RO. J Environ Manage 295:113133. https://doi.org/10.1016/j.jenvman.2021.113133
23. Fan L, Boshnakov K (2010) Neural-network-based water quality monitoring for wastewater treatment processes. In: 2010 Sixth International Conference on Natural Computation 5, 1746–1748. https://doi.org/10.1109/icnc.2010.5584378
24. Neifar M, Sghaier I, Guembri M, Chouchane H, Mosbah A, Ouzari HI, Jaouani A, Cherif A (2019). Recent advances in textile wastewater treatment using microbial consortia. J TextE Eng Fash Technol 5(3). https://doi.org/10.15406/jteft.2019.05.00194
25. Datta A, Philip L (2012) Biodegradation of volatile organic compounds from paint industries. Appl Biochem Biotechnol 167(3):564–580. https://doi.org/10.1007/s12010-012-9706-8
26. Liu Z, Lu Y, Wang P, Wang T, Liu S, Johnson AC, Sweetman AJ, Baninla Y (2017) Pollution pathways and release estimation of perfluorooctane sulfonate (PFOS) and perfluorooctanoic acid (PFOA) in central and eastern China. Sci Total Environ 580:1247–1256. https://doi.org/10.1016/j.scitotenv.2016.12.085
27. Lokhande S, RU, Singare P, Pimple S (2012). Study on physico-chemical parameters of waste water effluents from taloja industrial area of mumbai, india. Int J Ecosyst 1(1):1–9. https://doi.org/10.5923/j.ije.20110101.01
28. Wang Z, Xue M, Huang K, Liu Z (2011) Textile dyeing wastewater treatment. In Advances in Treating Textile Effluent. InTech. https://doi.org/10.5772/22670
29. Rajkumar D, Palanivelu K (2004) Electrochemical treatment of industrial wastewater. J Hazard Mater 113(1–3):123–129. https://doi.org/10.1016/j.jhazmat.2004.05.039
30. Rowdhwal SSS, Chen J (2018) Toxic effects of di-2-ethylhexyl phthalate: An overview. Biomed Res Int 2018:1–10. https://doi.org/10.1155/2018/1750368
31. Tang Z, Zhang L, Huang Q, Yang Y, Nie Z, Cheng J, Yang J, Wang Y, Chai M (2015) Contamination and risk of heavy metals in soils and sediments from a typical plastic waste recycling area in North China. Ecotoxicol Environ Saf 122:343–351. https://doi.org/10.1016/j.ecoenv.2015.08.006
32. Wu S, Wallace S, Brix H, Kuschk P, Kipkemoi K, Masi F, Dong R (2015) Treatment of industrial effluents in constructed wetlands: challenges, operational strategies and overall performance. Environmental Pollution. vol. 201 p. 107–120 (1) (PDF) The use of constructed wetlands for the treatment of industrial wastewater. Available from:
33. Kumar A, Arora PK (2022) Biotechnological applications of manganese peroxidases for sustainable management. Front Environ Sci 10. https://doi.org/10.3389/fenvs.2022.875157
34. Efrain Merma Chacca D, Maldonado I, Vilca FZ (2022) Environmental and ecotoxicological effects of drugs used for the treatment of COVID 19. Front Environ Sci 10. https://doi.org/10.3389/fenvs.2022.940975
35. Prakash D, Kumar R, Rajan K, Patel A, Yadav DK, Dhanker R, Khudsar FA (2022) Integrated application of macrophytes and zooplankton for wastewater treatment. Front Environ Sci 10. https://doi.org/10.3389/fenvs.2022.941841
36. Bera SP, Shah MP, Godhaniya M (2022) Microbial remediation of textile dye acid orange by a novel bacterial consortium SPB92. Front Environ Sci 10. https://doi.org/10.3389/fenvs.2022.930616
37. Deepa, Ameen F, Amirul Islam M, Dhanker R (2022) Green synthesis of silver nanoparticles from vegetable waste of pea Pisum sativum and bottle gourd Lagenaria siceraria: Characterization and antibacterial properties. Front Environ Sci 10. https://doi.org/10.3389/fenvs.2022.941554
38. Singh S, Bhardwaj AK, Dwivedi K (2024) Advances of plant-assisted synthesized metal nanoparticles and their environmental application. In: Biogenic Wastes-Enabled Nanomaterial Synthesis (pp. 373–392). Springer Nature Switzerland. https://doi.org/10.1007/978-3-031-59083-2_15
39. Mallick R, Rai S, Dwivedi K, Shukla N, Bhardwaj A (2023) Water resource management—a sustainable approach towards ground water conservation. In: Water Science and Technology Library (pp 179–188). Springer International Publishing. https://doi.org/10.1007/978-3-031-20208-7_11

40. Sonune A, Ghate R (2004) Developments in wastewater treatment methods. Desalination 167:55–63. https://doi.org/10.1016/j.desal.2004.06.113
41. Strecker E, Huber W, Heaney J, Bodine D, Sansalone J, Quigley M, Leisenring M, Pankani D, Thayumanavan A (2005) Critical Assessment of Stormwater Treatment and Control Selection Issues, 02 SW-1; Water Environment Research Foundation: Alexandria, Virginia
42. Daigger GT (2009) Evolving urban water and residuals management paradigms: Water reclamation and reuse, decentralization, and resource recovery. Water Environ Res 81(8):809–823. https://doi.org/10.2175/106143009x425898
43. Ojha N, Karn R, Abbas S, Bhugra S (2021) Bioremediation of industrial wastewater: A review. IOP Conf Ser: Earth Environ Sci 796(1):012012. https://doi.org/10.1088/1755-1315/796/1/012012
44. Shah MP (2018) Bioremediation-Waste water treatment. J Bioremediation Biodegrad 09(01). https://doi.org/10.4172/2155-6199.1000427
45. Sharma A, Mishra M, Sheet S, Thite M (2013) Role of microbes as cleaning degrading industrial wastes for environmental sustainability-A Reveiw. Recent Res Sci Technol 5(5)
46. Ali A, Naseem F (2012) Frequency distribution of bacteria isolated from different industrial effluents. Daffodil Int Univ J Sci Technol 7(1):28–33
47. Castellanos-Estupiñan M, Carrillo-Botello A, Rozo-Granados L, Becerra-Moreno D, García-Martínez J, Urbina-Suarez N, López-Barrera G, Barajas-Solano A, Bryan S, Zuorro A (2022) Removal of nutrients and pesticides from agricultural runoff using microalgae and cyanobacteria. Water 14(4):558. https://doi.org/10.3390/w14040558
48. Al-Jabri H, Das P, Khan S, Thaher M, AbdulQuadir M (2020) Treatment of wastewaters by microalgae and the potential applications of the produced biomass—a review. Water 13(1):27
49. Aghalari Z, Dahms HU, Sillanpää M, Sosa-Hernandez JE, Parra-Saldívar R (2020) Effectiveness of wastewater treatment systems in removing microbial agents: a systematic review. Global Health 16(1):13. https://doi.org/10.1186/s12992-020-0546-y, PMID 32013988
50. Qadir G (2019) Yeast a magical microorganism in the wastewater treatment. J Pharmacogn Phytochem. 8:1498–1500
51. Romdhana MH, Lecomte D, Ladevie B, Sablayrolles C (2009) Monitoring of pathogenic microorganisms' contamination during heat drying process of sewage sludge. Process Saf Environ Prot 87(6):377–386. https://doi.org/10.1016/j.psep.2009.08.003
52. Cabasso I (1987) Membranes in encyclopedia of polymer science and engineering, vol 9, pp 509–579. Wiley, New York

Urban Sustainability and Air Pollution Emission

Surbhi Narayan Lolage

Abstract Urban sustainability involves reducing and managing pollution emissions within cities to maintain environmental quality and public health. Urban areas often have high population densities, intricate transport network and dense commercial and industrial areas, which can compound existing environmental and health issue cities with their concentrated population and infrastructure are especially vulnerable to certain effects of climate change and other environmental stresses like heatweaves, flash flood, pollution, noise and soil contamination. In Europe, cities were exposed to levels of air pollution by fine particulate matters above the WHO guidelines. All these aspects have disproportionate impact on vulnerable segments of population and can exacerbate social challenges like segregation, poverty and inequality. Historically, municipalites have played crucial role in improving waste, wastewater and water management, public transport and land use via integrated urban planning approaches, CO2 emissions proxies to quantify urban CO2 emissions in many cities. China is an important intermediate case with rapid urban area growth combined with slower per capita CO2 emissions growth. Many cities also have taken leadership on greenhouse gas mitigation, pledging ambitious reduction targets. Pollution is result of economic growth and urbanization. Air pollution has been progressively recognized as serious problem for cities, through widespread effects on health and well being. Respiratory health disease and social inequality have increased due to air pollution and COVID-19 pandemic. Urban sustainability is crucial for mitigating the adverse impacts of rapid urbanization on environmental quality and public health. Strategies for sustainability include promoting renewable energy adoption, enhancing public transportation infrastructure, implementing green building practices.

Keywords Sustainability · Climate change · Population · CO2 emission

S. N. Lolage (✉)
Department of Microbiology, Padmasri Vikhe Patil College of Arts Science and Commerce, Pravaranagar, Maharashtra, India
e-mail: surbhilolage01@gmail.com

J. A. Parray et al. (eds.), *Sustainable Urban Environment and Waste Management*, Urban Sustainability, https://doi.org/10.1007/978-981-96-1140-9_5

1 Introduction

The Brundtland Report defined the concept of sustainable development as 'one that meets the needs of the present without compromising the ability of future generations'. Urban sustainability is not just about making towns and cities more efficient in terms of their use of resources. Air quality in urban areas can also depend on vehicular traffic, domestic heating, ship or emissions, energy production and others. Urban agglomerations and rural ecosystems in the Mediterranean region and globally are interlinked through the flow of resources, nutrients and wastes. Increases in diseases due to increase in pollutants in air cause economic and financial damage and social damage. Decrease in attractiveness of city due to high pollution rates can lead to economic and financial damage. The application of sustainable development is more advanced in cities in Northern Europe than in Mediterranean area. Our cities are naturally path-dependent due to their infrastructure and various historical circumtances [1], but they also have a pivotal role in global transition to sustainability [2].

Urban areas often have high population densities, intricate transport network and dense commercial and industrial areas, which can compound existing environmental and health issue. All these aspects have disaproportionate impact on vulnerable segments of population and can exacerbate social challenges like segregation, poverty and inequality. Respiratory health disease and social inequality have increased due to air pollution and COVID-19 pandemic. Urban sustainability is crucial for mitigating the adverse impacts of rapid urbanization on environmental quality and public health.

2 Urban Sustainability

According to Allen (1980), sustainable development means achieving enduring development addressing human needs and improvement of quality of life. Urban sustainability aims to improve the social, economic and environmental conditions of city to ensure the quality of life and well- being of current and future residents. Major sustainable urban development project include. Mixed-use development, transport-oriented development and green building construction. Achieving urban sustainable has been one of the main goals in European Union Policies since 1991.

About 75% of power generated globally is consumed in cities (1). Generating city-integrated energy at site of energy use could substantially contribute to environmental, economic and social aspect of urban sustainability since the early 2000s urban studies have focused on urban sustainability.

Changes in technological and social-ecological systems also represent an opportunity to prepare for sustainable and resilient urban transformation [3, 4]. The land area covered by cities covers about 5% of planet. Urban areas are characterized by high-density population, which accommodated by the development of extensive road networks, housing schemes, service and production industries and recreational

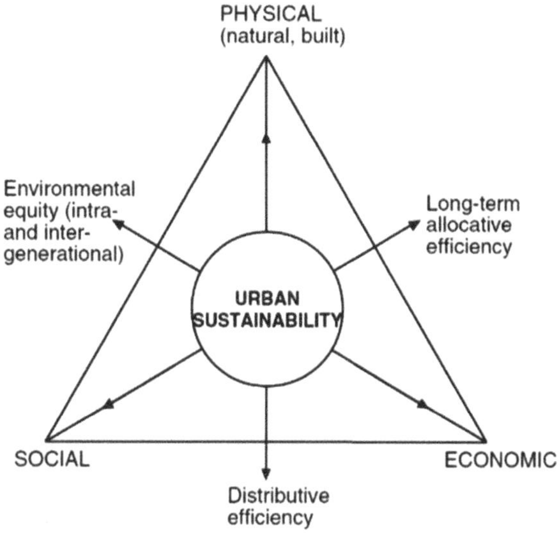

Fig. 1 Urban sustainability principles and policies (source)

facilities. Scientist directly affects the well being and happiness of urban inhabitants (Fig. 1).

3 Challenges

As cities continue to grow the urban environment faces particular sustainability challenges such as pollution, ecosystem and farmland protection and energy resources. Pollution is, by and large, the most common issue facing cities, specifically how to reduce air pollution, deal with waste and reduce water pollution.

Since urban environments are so densely populated, they are prone to greenhouse gas production, which causes air pollution. Greenhouse gas (GHG) production results from activities that burn fossil fuels. In addition, air pollution presents health risks to urban residents. Same as air pollution urban areas creates large water pollution problems. Debris, chemical and industrial runoff, and waste all pollute waterways, resulting in adverse effects on ecosystems and health of residents.

Providing clean water to residents is complex issues facing cities. A few large sustainability issues facing cities are ecosystem protection, food production and farmland protection. Urban farming or growing of small amounts of food often in fenced courtyards or on rooftops is becoming increasingly popular. About three-quarters of Europeans live in cities with their concentrated population and infrastructure is especially vulnerable to certain effects of climate change, heat waves, flashfloods, pollution, and noise and soil contamination. In 2020, 96% of people living in EU cities were exposed to levels of air pollution by fine particulate matter (PM2.5) above the WHO guideline. Technical, behavioral, economic and policy challenges have been identified as barriers to urban decarbonization. The downsides of current UWM system are its strong dependence on large quantities of water, high investment costs, and a need for stable institutions, long planning horizons and inefficient use of resources. Fossil fuels that produce such a large contributions to acid rain, global warming and urban air pollution.

4 Rate of Pollution Emission

Air pollution is usually concentrated in densely populated urban areas, especially in developing countries where the environmental regulations are relatively lack or non-existent. Therefore, urban air quality is a major concern throughout the world. Pollutants can be in the form of solid particles, liquid droplets or gases. Some air pollutants are poisonous and inhaling them can increase health problems. Pollutant can be classified as primary and secondary. Primary pollutants are directly produced from a process, such as ash from volcanic eruption sulphur dioxide released from factories. Secondary pollutants are form in air when primary pollutants react or interact. Polluted air can cause respiratory irritation or breathing difficulties during exercise or outdoor activities.

In most mega cities, the principle source of air pollution is motor vehicles. In some countries it have coal-fired power plants and other industrial facilities within city limits, so the levels of So2, NO_2 and particulates are correspondingly high (Masters and Ela 2008). Worldwide, WHO ranked Mexico city as the world's most polluted city. There are many sources of gases and particulate matter that pollute urban atmosphere. There have 6 criteria of pollutants, Carbon monoxide (CO), Sulfur dioxide (SO2), Nitrogen dioxide (NO_2), Tropospheric ozone (O3), Particulate matter, Lead.

5 Carbon Monoxide (CO)

It is colourless and odourless gas. Produced by incomplete combustion of fuel. Vehicles and engines account for 80% of CO emissions. It is harmful for living being cause it can bind irreversibly to hemoglobin in red blood cells, preventing the hemoglobin from binding with oxygen.

6 Sulfur Dioxide (SO2)

It is a colourless gas has pungent odor. Majority of So2 pollution results from the combustion of coal for electricity generation and industry. In atmosphere So2 may react to form sulfur trioxide (SO3) and sulfuric acid (H2SO4) which then settle back to Earth in form of acid depositiom.

7 Nitrogen Dioxide (NO2)

It is highly reactive, foul-smelling reddish-brown gas. It contributes to smog and acid deposition. Most of NO2 emissions result from combustion in vehicle engine, industrial combustion.

8 Tropospheric Ozone (O3)

Ozone is colourless gas with an objectionable odour results from the interaction of sunlight, heat, nitrogen oxides and volatile carbon-containing chemicals. When ozone comes into contact with surface it rapidly releases this extra force in the form of chemical energy. It can cause respiratory tract diseases.

9 Particulate Matter

Particulate matter includes primary pollutants such as dust and soot. Secondary pollutant such as sulfates and nitrates. Most PM10 pollution is dust whereas most PM2.5 pollution results from combustion processes. It can damage respiratory tissues when inhaled.

10 Lead

Lead is emitted into atmosphere primarily in the form of inorganic particulates. Airborne lead may effect human population by direct inhalation. Lead can leach out of those older water system especially if the water is acidic or particularly soft.

Over the past several years number of disease has increased greatly such as Asthma, allergic reactions, bronchitis and respiratory infection. The effects of air pollution on health are dependent on several factors like age, general state of health, duration of exposure. Therefore, air pollution is major environmental health issues.

Air pollution also causes damage to plants and animals, affecting biodiversity and crop yields. Exposure to air pollutants is largely beyond the control of individuals and requires action by public authorities at national, regional and even international levels. Heavy-duty trucks (HDTs) are major contributors to urban traffic pollution. In China, HDTs emits 16.8% of carbon monoxide (CO), 6.9% of total hydrocarbon (THC), 57.8% of nitrogen oxides (NOx) and 66.3% of particulate matter (PM) of the total motor vehicle emissions, despite only constituting 3.1% of vehicle population. Since 2007, global population has been increased in living urban areas than in rural areas.

It is estimated that cities are currently responsible for 60–70% of greenhouse gas emissions. The energy systems of cities are increasingly vulnerable to effects of climate change and extreme weather such as storms, flooding and sea-level rise and also to natural and human-induced disasters. Urban energy systems directly affect the well-being and happiness of urban inhabitants. About 75% of power generated globally is consumed in cities. It was observed that fuel consumption rapidly increases owing to waste collection increase. Number of vehicles, their types and usage significantly influence fuel consumption. BOD and COD concentrations increase with Municipal Solid Waste (MSW) collection coverage and landfill rate due to increasing amount of wastewater generated by higher quantity of waste landfilled. More than half of the world's population depends on traditional biomass or other energy sources that result in unhealthy indoor air pollution and the International Energy Agency expects this trend to continue to rise. The use of traditional biomass fuels for cooking and heating arises from complicated set of factors ranging from cultural tradition and behavior to the burdens and limits of poverty (Table 1).

11 Case Studies

In Africa, annually 125 million tons of Municipal Solid Waste (MSW) is generated from which only 55% is collected and in Sub-Saharan Africa 44%. In between 2000 and 2015 African cities has been growing rapidly with 150%. This waste site increases the occurance of diseases like malaria, cholera, typhoid, yellow fever, dengue, hepatitis because environmental condition is ideal for the breeding of mosquitoes, flies and pathogens. Almost 23% of vector-borne, diarrheal and cardiovascular diseases are attributed to environmental factors in Africa. The weak MSW management system also affects the economy and social life as it can lead to lack of motivation and demoralized mindset.

Waste management in some European and American countries became more organized in administration and techniques, by related policies and investment in infrastructure after the industrial revolution. The city of Kinshasa has demonstrated efforts to ameliorate the current MSW situation by creating waste management campaigns name 'Kin bopeto', while also getting involved in international platforms on MSW such as African Clean Cities Platforms (ACCP). The research focuses on

Table 1 Major air pollutants (*Source* Environmental protection agency)

Pollutant	Composition	Primary or secondary	Characteristics
Particulate matter			
Dust	Variable	Primary	Solid particles
Lead	Pb	Primary	Solid particles
Sulfuric acid	H_2SO_4	Secondary	Liquid droplets
Nitrogen oxides			
Nitrogen dioxide	NO_2	Primary	Reddish-brown gas
Sulfur oxides			
Sulfur dioxide	SO_2	Primary	Colorless gas with strong odor
Carbon oxides			
Carbon monoxide	CO	Primary	Colorless, odorless gas
Carbon dioxide	CO_2	Primary	Colorless, odorless gas
Hydrocarbons			
Methane	CH_4	Primary	Colorless, odorless gas
Benzene	C_6H_6	Primary	Liquid with sweet smell
Ozone	O_3	Secondary	Pale blue gas with acrid odor
Air toxics			
Chlorine	Cl_2	Primary	Yellow-green gas

comparing parameters such as management cost, GHG emissions, waste pollution, fuel consumption and landfill diversion.

Many African countries have undertaken actions enhancing their solid waste management systems. Kenya has increased waste recycling by 90% with Taka project which is a private company that has helped to enable affordable waste collection, sorting, composting, plastic recycling and waste purchasing in poor areas of country.

Municipal Solid Waste management system in Kinshasa city, by increasing waste collection coverage and implementing diverse streams of waste valorization, causes a decrease of atmospheric pollution in terms of CO_2 equivalent by half, when all the waste is collected, and a landfill diversion increase up to 70%. Urban air pollution of United States has failed to meet Clean Air Act requirement governing the scope and timing of reductions. Using five years data (2002–2006), emissions of carbon monoxide declined by 10%, volatile organic compound by 8%, nitrogen oxide by 14% and hazardous air pollutants for sources reporting to toxic release inventory by 8% (2001–2005). The United States continues to rely predominantly upon coal-fired power production, a very dirty energy source contributing to acid rain, climate change and urban pollution, just as it did prior to Earth Summit.

California's LEVII program introduced the hybrid vehicles, this program demands significant emission reductions that effectively force significant technological change. WLPGA/ UNDP LP Gas Rural Energy Challenge in South Africa creates commercially sustainable gas markets for low-income household and reducing health and environmental impacts. Improved health as result of reduced indoor air pollution is one of the benefits of using liquefied petroleum (LP) gas rather than other fuels like firewood or coal, in a domestic environment.

A community science campaign (# CHEARIATIRA) was carried out in February 2019 in Torino (Italy), with the aim of engaging the public in measuring NO2 concentrations in an urban area that often exceeds air quality standards.

In Naples, Italy Vehicular traffic emissions can significantly jeopardize air quality in historical centers, as demonstrated by a monitoring campaign of fine particles (FPs) ranging in size from 20 to 1000 nm.

In the United States, the urban proportion of the population has been over 80% of the last decades, it is projected to reach about 89% by 2050. During the same period, China will experience even more rapid urbanization; its urban population will increase from about 50% to about 78%. China will be the first country with 1 billion urban occupants by 2050. This rapid expansion in cities in both the United State and China has and will continue to place pressure on environmental, economic and social systems.

In December 2019, the New Crown Pneumonia (the COVID-19) out broke around the globe, and China imposed a nationwide lockdown starting as early as January 23, 2020. This decision has significantly impacted China's air quality, especially the sharp decrease in PM2.5 pollution.

12 Management Directions

Methods by which cities are seeking to reduce air pollution include alternative forms of transportation such as walking, biking, public transportation. Some policies to regulate business and increasing green space. Sanitation can either reduce pollution or exacerbate it, depending on how it is done, such as improving waste collection, creating and promoting recycling programs, implementing and promoting composting programs and public education regarding waste management and sanitation. For urban planning improving the energy efficiency of building and making cities more green and compact. Ensuring access to affordable clean household energy solutions for cooking, heating and lighting. Increased use of low-emissions fuels and renewable combustion free power sources like solar, wind and hydropower.

The main two strategies for transitioning to low-carbon cities are to shift from fossil fuels to cleaner energy sources and to reduce urban energy consumption levels. The low-carbon transition can be accomplished through energy efficiency measures, behavioral interventions and incorporating carbon sinks such as urban parks. Life Cycle Assessment (LCA) is an input–output method to determine the environmental impact of materials creation usage and end of life. It can be applied to product such as plastic bottle or waste management. Using LCA, it has been determined that recycling and other methods of waste management improvement can reduce up to 15% of global Greenhouse Gas (GHG) emissions (Fig. 2).

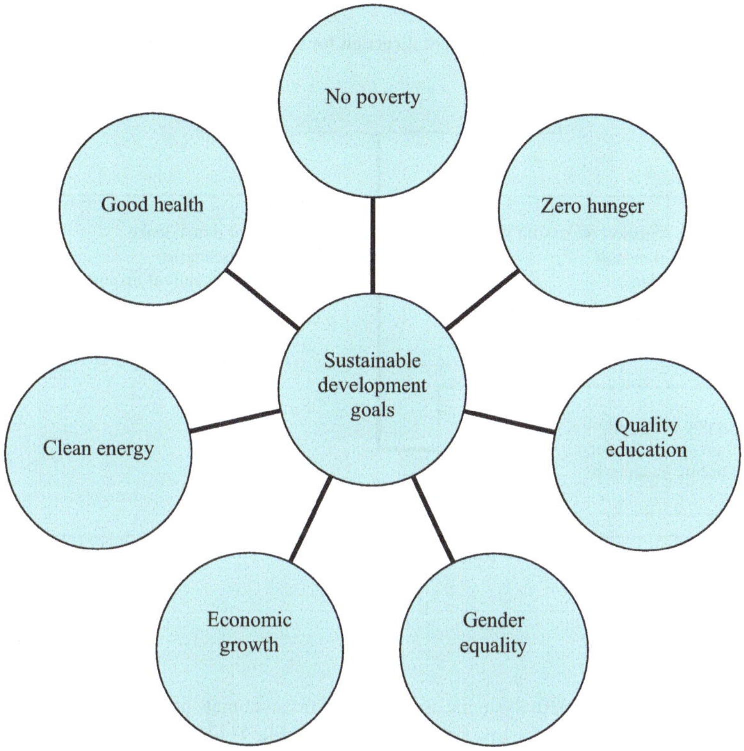

Fig. 2 Over view of SDG goals

Lock-in is path-dependent process that can result in future transformations being prohibited due to set of favorable initial conditions and benefits? To avoid long-term harmful consequences, mainstreaming lock-in assessments should be a future research direction of urban studies. Urban wind energy provides opportunities for renewable electrical generation but also ventilation, pollution dispersion and seriousness of urban heat-island effect.

To limit global climate change to 2 degree Celsius above preindustrial levels, the greatest global investment is required in the building sector. Many city governments encourage energy storage and low-carbon generation at the building level through smart net-metered urban distribution networks. The top priorities for urban water sustainability include the provision of safe drinking water, wastewater handling for the public health and protection against flooding. The UWM (Urban Water Management) system relies on investment-intensive, usually underground, pipe networks that provide single quality drinking water and evacuate storm water and wastewater. The two main strategies for transitioning to a low-carbon city are to shift from fossil fuels to cleaner energy sources and to reduce urban energy consumption levels. The United States has generally made substantial progress towards sustainable development goals for stratospheric ozone depletion (Fig. 3).

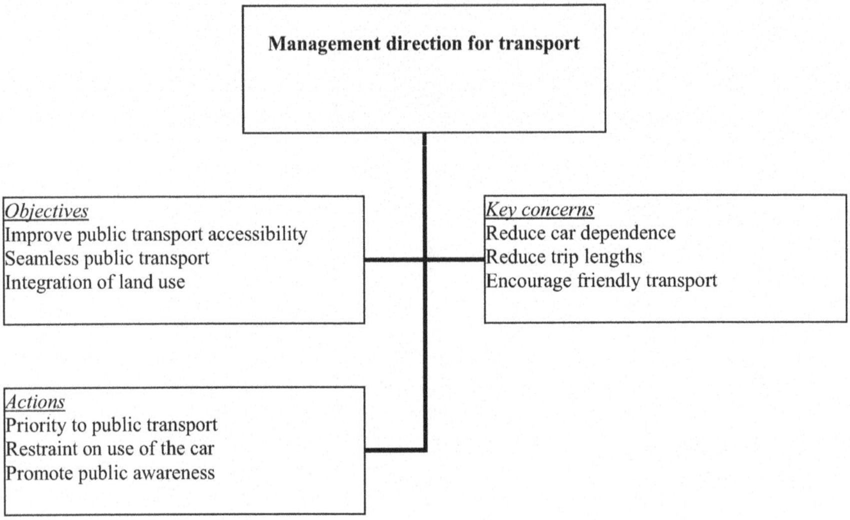

Fig. 3 Management directions for transport

References

1. van den Bergh JCJM (2020) Systemic assessment of urban climate policies worldwide: decomposing effectiveness into 3 factors. Environ Sci Policy 114:35–42. https://doi.org/10.1016/j.envsci.2020.07.011
2. Wolfram M (2016) Conceptualizing urban transformative capacity: a framework for research and policy. Cities 51:121–130. https://doi.org/10.1016/j.cities.2015.11.011
3. Amirzadeh M, Sobhaninia S, Sharifi A (2022) Urban resilience: a vague or an evolutionary concept? Sustain Cities Soc 81:103853. https://doi.org/10.1016/j.scs.2022.103853
4. Egerer M, Haase D, McPhearson T, Frantzeskaki N, Andersson E, Nagendra H, Ossola A (2021) Urban change as an untapped opportunity for climate adaptation. npj Urban Sustain. https://doi.org/10.1038/s42949-021-00024-y

Innovative Waste Treatment Techniques: Urban Sustainability Challenges

Bisma Rashid, Zaffar Bashir, Marifa Gulzar, Nusrat Shafi, and Javid Ahmad Parray

Abstract Waste and humans along with other organisms have a long history together. It comes from a variety of sources, including agricultural crops leftovers, the food supply chain and its production, and sectors and their activities. Waste can contaminate groundwater and has various negative effects on the environment, such as leachate. Pollution occurs in the air due to emissions from burning waste, and in the soil due to its disposal. The proper handling of garbage and trash has emerged as a crucial obstacle to sustainable urban growth as urban residents continue to rise. There are serious threats to the environment, human health, and the economy associated with the growing amount and complexity of trash produced by cities. To reduce these dangers, efficient solid waste treatment methods are necessary. Various technologies and methods are employed globally to reduce waste.

Keywords Waste management · Sustainable waste management · Composting · Land fillin

B. Rashid (✉)
Division of Basic Sciences and Humanities, Faculty of Agriculture, SKUAST-Kashmir, Wadura 193201, India
e-mail: bismarashidbhat@gmail.com

Z. Bashir
Centre of Research for Development and PG Programme in Microbiology, University of Kashmir, Srinagar 190006, India

M. Gulzar
Division of Horticulture, SKUAST-Kashmir, Srinagar 190025, India

N. Shafi
Department of Chemistry, Government Degree College Eidgah, Kashmir, Srinagar 190011, India

J. A. Parray
Department of Environmental Science, Government Degree College Eidgah, Kashmir, Srinagar 190011, India

J. A. Parray et al. (eds.), *Sustainable Urban Environment and Waste Management*, Urban Sustainability, https://doi.org/10.1007/978-981-96-1140-9_6

1 Introduction

Any solid material that originates from humans or other living things and is deemed undesirable is referred to as solid waste. Solid wastes come in both organic and inorganic forms, and they are typically worthless to living things. Kitchen trash, wrapping, newspaper articles, cloth, plastic bottles, cans, newspapers, and dry batteries are examples of wastes. Based on factors like biodegradability, inertness, e-waste, composites, and domesticity, there are various categories of solid waste. Both household and industrial activities produce all of these wastes [1]. Through approximately 2 billion tons of solid waste from municipalities generated annually, the world is facing an increasing solid waste management crisis. Urban sustainability is facing major obstacles as a result of the growing strain on the infrastructure for waste management brought on by growing urban populations. To reduce the negative effects of waste disposal on the environment, society, and economy, efficient treatment of solid waste technologies must be developed and put into practice. Landfills are a major component of many cities' current waste management strategies, which can result in contamination, emissions of greenhouse gases, and contamination of groundwater and soil. Furthermore, the environmental issues surrounding waste disposal are made worse by the mining of natural resources needed to produce packaging materials like plastics. As a result, it is imperative to implement creative and sustainable waste management techniques that reduce the environmental impact of waste production and disposal. Adopting cutting-edge garbage treatment technologies, like composting, recycling, and energy-from-waste facilities, is one of the most promising solutions. These innovations have the potential to produce renewable energy, preserve natural resources, and drastically cut down on the quantity of waste dumped in landfills. Recycling, for example, can cut the need for virgin materials by up to 70%, and energy-from-waste facilities can use waste that cannot be recycled to produce heat and electricity. Composting also has the ability to transform organic waste into beneficial soil amendments and fertilizers. But putting these technologies into practice calls for a large investment in technology and infrastructure, as well as modifications to public attitudes and behaviour regarding waste disposal. To ensure successful implementation, cities must also create comprehensive waste management plans that incorporate a variety of technologies and stakeholder engagement. Additionally, regulatory frameworks that facilitate the implementation of environmentally friendly waste management procedures and promote investment from the private sector in infrastructure for waste management must be provided by governments. Urban handling solid waste presents a number of challenges that call for a multifaceted strategy that includes stakeholder engagement, policy changes, and cutting edge technologies. Cities can lessen their environmental impact, preserve natural resources, and foster economic growth through the implementation of environmentally friendly waste management procedures and implementing cutting-edge solid waste treatment technologies. In order to address this urgent global issue, governments, businesses in the private sector, and organizations representing civil society must work together to develop practical solutions for urban sustainability. The rate at which metropolitan waste is

Table 1 Annual regional waste generation

Region	Waste generation (Million Tonnes)
East Asia & The Pacific	468
Europe & Central Asia	392
South Asia	334
North America	289
Latin America & The Caribbean	231
Sub-Saharan Africa	174
Middle East & North Africa	129

Sourcesand types of solid wastes

generated in developing countries has dramatically increased due to factors such as population growth, economic expansion, and rapid urbanization [2]. Waste generation is usually the responsibility of cities. The risks associated with solid waste materials and the technologies used to treat them vary. Garbage that offends people and causes them discomfort wherever it is located corresponds to the way leachates from landfills taint the groundwater and soil while greenhouse emissions of gases via combustion, incineration, pyrolysis, and waste-to-energy operations create pollutants in the atmosphere and global warming [3] (Table 1).

The following is a brief description of the various solid waste types that escape from various sources.

2 Biodegradable Trash

This type of waste includes any organic waste that microorganisms and other organisms can break down into simple organic molecules, water, carbon dioxide, or methane. Both aerobic and anaerobic digestion as well as composting is used to break down these wastes. These wastes are broken down by the aforementioned processes both domestically and commercially in order to reduce waste volumes, lessen negative effects on the health of humans and the environment, and produce useful products that can be used wherever possible. Correct selection of the various biodegradable waste compositions and maintenance of an appropriate temperature regime are crucial. Usually, biodegradable waste is buried, burned, or damped into the ground. Because organic wastes degrade slowly, they release gases into the atmosphere that cannot be captured and add to the greenhouse gas emissions. About 30% of the world's methane emissions are thought to come from this process [4]. The synthesis of carbon-based adsorbents from biodegradable wastes has emerged as a very intriguing substitute for burning, recycling, or composting in order to reduce the release of gases emission levels from the wastes' breakdown [5]. Activated carbon is a great desiccant for cleansing gases and liquids because of its extensive surface area and micro-porous framework, which allow it to have excellent sorption ability regarding an extensive variety of gases and other compounds. The ability to create activated carbon from any waste product containing carbon is an additional advantage. Because of the expensive expense of creating activated carbon from fossil coals, as well as other ecological and economic concerns, the use of organic materials for this purpose has gained a lot of interest recently. Adsorption tests have been carried out on activated carbons obtained from post-agricultural manufacturing waste in both gas and liquid phases due to its affordability, availability, and renewable nature.

3 Reusable

Recyclable waste is simply any waste that can be repurposed for something positive. It is the process of repurposing waste materials in their entirety or transforming them into new, useful materials and objects. It is the greatest substitute for traditional waste disposal since it reduces greenhouse gas emissions and preserves natural resources. It can cut down on the usage of fresh raw materials and avoid wasting potentially useful resources and materials. This lowers energy consumption as well as air and water pollution from land filling and incineration. Recycling is a key element of the modern three "Rs" (Reduce, reuse, and recycle) waste management approach. Thus, by replacing raw material inputs, it strives for environmental sustainability. Glass, cardboard, paper, aluminium, plastics, metals, tires, batteries, and electronic equipment are among the materials that can be recycled. These substances are either transported to a recycling facility where they are sorted, cleaned, and processed once again to create new materials that will be used to make new goods. The world's problems with resource depletion, climate change, and environmental pollution have gotten worse due to the rapid growth of the economy and population. In order to address the aforementioned problems concurrently, recyclable waste disposal is becoming increasingly important, and in recent years, governments have started to control and encourage separation of sources and urban waste recycling. Mainstream, secondary, and tertiary recycling are the three primary categories. In primary recycling, waste items are transformed into similar resources, including newspapers into more paper. Wastes are repurposed in the subsequent recycling process, including cutting up bottles of plastic for planting in the garden. Wastes are entirely converted into novel goods via the ultimate recycling process using burning, chemical reactions, or other techniques.

4 Inert

Any kind of waste that does not break down and cannot be broken down chemically or biologically. It makes up the majority of waste and can account for up to 50% of all waste produced, making management difficult. Because it contains a variety of secondary pollutants that are trapped in it, it also significantly affects both human health and the environment. Demolition waste, rubber, plastic, ceramic, and other debris are examples of inert wastes. Its management only includes recycling it as base material and filler for surfaces, concrete aggregate, building materials, and tiles, among other uses.

5 Electronic Waste

It is electrical device waste that has been abandoned. Refurbished electronics can be sold again, used again, or their mother's circuit boards and metals can be recovered and reused in new equipment. Improper management of e-waste can have detrimental effects on the environment and public health in developing countries. This is due to the possibility that electronic waste, such as CPUs, may contain potentially dangerous elements like flame retardants based on bromine, cadmium, lead, or beryllium. Recycling and getting rid of e-waste can pose serious risks to the health of workers and the communities in which they are employed. Ten categories—small appliances, large household appliances (such as freezers and coolers), for the home (gadgets), electronics for consumers (television), lamps, toys, tools, healthcare equipment, control and surveillance instruments, and automatic dispensers—are used by the Electronic Equipment Directive to classify e-waste. Nonetheless, some people divide e-waste into six groups: (1) temperature-exchange devices (freezers, air conditioners), (2) screens and cameras, (3) illumination, (4) large appliances (electric stoves, washing machines), (5) small appliances (electric shaver, microwave), and (6) minor IT and communication devices (printers, cell phones). Among other distinctions, products within each category differ in terms of impact, longevity profile, and collection techniques. The largest volume of e-waste (18.2 Mt) was imported into Asia in 2016, with Europe (12.3 Mt), America (11.3 Mt), Africa (2.2 Mt), and Oceania (0.7 Mt) following closely behind. Oceania, the least when it comes of total e-waste produced, generated the most e-waste per person (17.3 kg), with only 6% of e-waste being collected and recycled. Through a mean of 16.6 kg of e-waste generated per person, Europe is the second-largest producer.

6 Composite

An amalgam of more than one substance is called a composite. Dealing with multiple types of waste at once is difficult, which leads to a lot of issues with this type of waste. Since there haven't been effective endeavors to reuse the components of composites up to this point, the trash is increasing every day. Composite waste includes liquid garbage, organic waste, solid and recyclable waste, and hazardous waste. Over the past ten years, researchers have focused on composite recycling, and several techniques have been put forth, including chemical recycling, the oxidation process in a bed of fluid, and pyrolysis. Of them, pyrolysis is the only one that is being used on a commercial scale at the moment.

7 Household Wastes

Household waste, sometimes referred to as domestic waste, is the garbage and debris produced by homes and residential areas. Both organic and inorganic items, such as food waste, paper goods, plastics, glass, and textiles, are commonly found in this kind of garbage. Numerous activities, including everyday living routines, housekeeping, and the acquisition of goods and services, can result in the generation of domestic trash. Depending on variables including location, way of life, and cultural customs, the makeup of household garbage might differ greatly. Families in rural locations might produce more organic trash from farming and raising livestock, whereas families in metropolitan areas may produce more packaging garbage from takeout food and online shopping. Furthermore, residences with pets may produce more pet waste, while those with big families or young children might produce more nappies and baby wipes. If not correctly managed, household wastes can have serious negative effects on the environment and human health. Household garbage that is not properly disposed of can contaminate water, air, and soil resources, leading to contamination and environmental deterioration. In addition, improper waste management can result in unsightly smells, the spread of illnesses and pests, and bad aesthetics. For example, uncollected waste or overflowing trash cans can draw insects, flies, rodents, and other insects that can harbor disease. Adopting efficient waste management strategies is crucial to reducing the negative effects that home wastes have on the environment and human health. This entails composting organic trash, separating waste into reusable components, and lowering waste production through sustainable consumption habits. Additionally, households can take part in activities for the collecting of bulk waste or in community recycling programs. Governments can also enact laws and rules to encourage appropriate waste management and disposal procedures. Home wastes are a serious environmental issue that has to be addressed. Residents may help contribute to an environmentally friendly and healthier world by implementing sustainable waste management techniques and minimizing trash output through mindful consumption behaviors. Governments may also be very important by putting laws and policies in place that support ethical waste management and disposal procedures. In the end, it is critical to acknowledge the significance of managing household garbage and to collaborate in order to lessen the detrimental effects that household waste has on our communities and the environment. Large volumes of municipal waste are being generated as a result of rapid urbanization. The environment becomes more polluted as municipal solid waste increases, and this needs to be managed appropriately. Therefore, there needs to be a method for disposing of solid waste from municipalities that works [6]. Using them as fertilizer and soil conditioner, or combining them with other wastes to make fuel, are some methods of getting rid of them. Additionally, these wastes are processed to produce the compost [7], utilized as bio sorbents for heavy metal removal [8], employed as bio-control substances, and utilized to generate electricity.

8 E-Wastes

E-waste, or electronic garbage, is a problem that is becoming more and more widespread. Electronic items that are either not anymore in use or have approached the end of their useful life, including phones, computers, televisions, and other products, are referred to as e-waste. An enormous rise in the production of electronic trash has resulted from the quick development of technology; estimates place the annual global production of e-waste at around 50 million metric tons. Consumer devices like tablets, computers, and cell phones—which are regularly swapped out for newer models—are the main sources of e-waste. To make matters worse, the discarding of obsolete gadgets like VHS players and tape recorders adds to the growing problem of e-waste. Because e-waste can contain hazardous materials like mercury, cadmium, and lead that can contaminate land and water if improperly disposed of, it has a considerable negative influence on the environment. Because many electronic gadgets include dangerous materials that need to be handled and disposed of carefully, managing e-waste is a complicated problem. Ineffective disposal techniques can lead to resource loss, worker and community health problems, and environmental contamination. To lessen the impact on the environment and recover precious elements like gold, copper, and silver, e-waste recycling is essential. The management of e-waste presents a variety of difficulties. Numerous nations lack sufficient facilities and policies to efficiently handle e-waste, resulting to illicit disposal and combustion of electronic gadgets. Furthermore, efficient waste management techniques are hampered by a lack of knowledge about e-waste problems and their effects on the environment. More attention has been paid in recent years to sustainable e-waste management techniques. To promote appropriate e-waste disposal and recycling, organizations and governments are putting rules and initiatives into place. For example, extended producer responsibility (EPR) laws have been implemented in certain nations, requiring producers to be accountable for the trash their products produce. Individuals may make a big difference in the growing e-waste problem by implementing responsible e-waste disposal methods. This involves recycling them at approved facilities or giving working electronic equipment to nonprofits. In order to lessen the production of electronic trash, consumers can also select products with ecologically friendly packaging and design. Eventually, governments, businesses, and individuals must work together to solve the e-waste issue. We can lessen the environmental effect of the generation of electronic trash and build a more sustainable future for future generations by supporting sustainable e-waste management techniques and ethical disposal behaviors. One of the waste streams that are expanding globally is electronic waste, or "E-waste." It has various potentially harmful effects on the environment. Piles of e-waste give the area a negative reputation, and heavy metals in it pollute the land and water. 2014 saw the production of 41 million metric

Fig. 1 Representative process diagram for solid waste handling

tons of e-waste worldwide. It is growing yearly at a pace of three to five percent. This makes it difficult for the researchers to come up with an appropriate solution to deal with the issue. Recycling is a tactic that has been employed for ages (Fig. 1).

9 Various Waste Management Innovations' Cost–benefit Ratios

9.1 Fuel Production

In order to create different kinds of fuels, raw materials must be extracted, processed, and refined. This is a critical step in the fuel manufacturing process. Fossil fuels—coal; oil, and natural gas—are the primary means of fuel production and have long served as the foundation of the world's energy industry. However, alternative fuels like hydrogen, biofuels, and nuclear energy are becoming more and more well-liked due to growing worries concerning global warming and environmental sustainability.

The mining and processing of essential elements from the soil is usually the first step in the fuel generation process. When it comes to fossil fuels, this entails extracting coal via open-pit mines or underground mines or extracting and pouring crude oil from subterranean reserves. The extraction of natural gas from subterranean formations follows a similar procedure. After the extraction process, the unprocessed products are shipped to processing centers or refineries to undergo additional processing. Refining is an essential stage in the fuel production process that creates different kinds of fuels by sorting and cleaning the raw ingredients. For instance, various petroleum products including gasoline, diesel fuel, aviation fuel, and lubricants are made from refined crude oil. Liquefied petroleum gas, also known as LPG, and other refined products are made from natural gas. Biofuels are created by processing biomass—such as sugarcane or maize starch—into biodiesel or ethanol. Fuel production includes not only refining but also processing non-traditional fuels such as nuclear energy and hydrogen. Hydrogen is a clean-burning fuel used in power generation and transportation that is created via electrolysis or steaming methane reforming. In a nuclear power plant, uranium atoms fission produces nuclear energy, which is utilized to create electricity. Fossil fuels dominate the worldwide fuel production landscape, providing approximately 80% of the globe's energy consumption. However, due to worries regarding global warming and environmental sustainability, there is an increasing need for alternative fuels. Investments in hydrogen and biofuels, as well as sources of clean energy like wind and solar power, have increased significantly in recent years. Fossil fuels will remain an important part of the world's energy balance for many years to come, despite the growing desire for alternative fuels. But as the globe moves toward a low-carbon economy, there's an increasing demand for cleaner, more sustainable fuel generation techniques. This includes the creation of innovative technologies that can boost productivity and lessen the negative effects of mining for fossil fuels on the environment. In order to create different kinds of fuels, raw materials must be extracted, processed, and refined—a difficult process. The need for renewable energies like biofuels, the gas hydrogen, and nuclear power is rising, even though fossil fuels will still make up the majority of the world's energy mix for many years to come. In order to meet our expanding energy needs while reducing our environmental effect, we must create more ecologically friendly and sustainable fuel production methods as the globe moves toward a low-carbon economy. Nowadays, many people are considering energy production from various solid wastes as a sustainable way to produce energy and dispose of waste at the same time. Because the feedstock in this method is typically waste, which would otherwise be a headache, it is incredibly cost-effective. Waste to energy has more advantages than energy from fossil fuels in terms of lowering greenhouse gas emissions and preventing global warming, according to a cost–benefit analysis. Cooking oil waste is converted into energy and used to make biodiesel. The industry of land filling and energy recovery from waste is advantageous.

10 Recycling

Gathering and preparing items that might otherwise be put away as waste is an aspect of recycling. This comprises materials that can be used to create novel products, including plastic, glass, paper, glass, and metallic material, which lessens the need to extract and process raw resources. Because recycling preserves natural resources, diminishes the negative consequences of human activities on the environment, and minimizes the amount of rubbish that ends up in landfills, it has become a necessary habit in today's world. One of recycling's biggest benefits is the ongoing preservation of natural resources. Recycling reduces the quantity of raw materials extracted from the earth, hence mitigating the negative impacts of mining, logging, and drilling on the ecosystem. For example, one ton of recycled paper saves 17 trees, 7,000 gallons of water, and 4,100 kilowatt-hours of electricity. In a comparable manner recovering a metric ton of aluminum cans generates 14,000 kilowatt-hours of power and 20 barrels of oil. Recycling contributes to the mitigation of climate change by lowering greenhouse gas emissions. Energy is needed to create innovative goods from raw materials, and this energy is often produced by charring fossil fuels and emitting greenhouse gases into the environment. Recycling lessens the requirement for this energy use and contributes to a reduction in the atmospheric concentration of carbon dioxide. Recycling can actually cut the release of greenhouse gases with upwards of 117 million tons of carbon dioxide annually in the US alone, according to studies. Recycling not only helps the environment, but it also benefits the economy. Recycling generates new industries and business opportunities, which boost economic growth and create jobs. Approximately 1.5 million people are employed in the recycling sector in the US, which also produces approximately $117 billion in annual economic activity. Recycling also contributes to the preservation of landfill space and lowers trash disposal expenses, both of which are important for businesses and governments. Effective recycling requires careful material preparation and sorting. Usually, this entails sorting recyclables from waste and impurities like liquids and food waste. Curbside recycling systems have been introduced in many areas, making it simple for locals to recycle. Additionally, some towns offer recycling drop-off locations or organize special bulk item collections. Recycling is an essential activity that lowers trash disposal costs, conserves natural resources, and lessens the negative effects of human activity on the environment. Communities may guarantee that recyclables are converted into new goods and lessen the need for mining and processing raw materials by correctly sorting and preparing materials. Together, individuals and governments can encourage recycling and minimize waste, paving the way for a future that is healthier for future generations. Reusing or selling used or scrap materials to areas that need them is a win–win situation known as recycling. The advantages outweigh the disadvantages [9]. Recycling allows for the repeated use of various resource types. Developed nations donate their outdated equipment to developing nations One way to recycle e-waste is to adopt new technology while donating outdated technology to developing nations. For instance, developing nations purchase outdated equipment at reduced costs. By using this approach, they are able to dispose of their electronic waste and sell it for a profit [10].

11 Future Statements of Study

Cities' waste management practices are changing as a result of urbanization, since rising population and changing consumer habits produce large volumes of municipal solid waste (MSW). The requirement for efficient and environmentally friendly waste management techniques grows as cities expand. Many cities still handle their trash using conventional techniques like recycling, incineration, and landfilling. These methods, however, have drawbacks, such as significant negative effects on the environment and society, and they are frequently not long-term viable. Innovative approaches to waste management will be necessary to meet future difficulties related to urban sustainability. In order to minimize landfilling and incineration and to encourage trash reduction, recycling, and reuse, cities must embrace the concepts of the circular economy. Utilizing cutting-edge technologies, such waste-to-energy (WTE) systems, is one viable option. By turning waste that cannot be recycled into energy, these systems can lessen their reliance on fossil fuels and greenhouse gas emissions. The creation of intelligent waste management systems is a crucial field of innovation. These systems optimize garbage collection routes, lower waste volumes, and boost recycling rates through the use of machine learning algorithms, IoT sensors, and data analytics. In order to get citizens involved in recycling and waste reduction initiatives, cities must also give priority to outreach and education initiatives. This entails encouraging actions like cutting back on plastic consumption, composting food scraps, and recycling properly. The next phase of metropolitan sustainability will involve tight coordination amongst government departments, private sector enterprises, and community organizations in order to develop successful waste management plans. This entails investing in R&D, coordinating policy initiatives, and exchanging best practices. Cities need to give infrastructure development top priority in addition to technological and legislative advancements in order to facilitate sustainable garbage management. This entails creating new recycling facilities, improving collection routes, and creating green areas for garbage processing. Target 12: Sustainable Production and Consumption intends to reduce the production of municipal solid waste per capita by 10% and cut global food waste per capita by half by 2030. Cities need to make the shift to zero-waste economies by 2050, with a focus on circular economy ideas, environmentally friendly production methods, and effective trash disposal techniques. In order to guarantee an environmentally friendly future for urban people, this calls for a sustained commitment to innovation, cooperation, and the creation of policies. An inventive method for managing solid wastes is sustainable solid waste management, which involves gathering, processing, and repurposing waste materials to produce fuel and energy. However, because combustion, pyrolysis, and other processes can pollute the air, investigations should be done to develop new fuels or modify existing methods of producing energy without endangering the already polluted air from excessive emissions of carbon and other greenhouse gases. The leaching of metallic substances and other contaminants from landfills contaminates the soil and groundwater. In order to reduce these risks, all of these concerns must be taken care of, along with the production of fuel and energy, and the ongoing and prudent utilization regarding resources accessible to environmental health issues and improvement.

12 Conclusion

In order to handle the complexity of garbage generation and treatment, creative solutions are needed in the urgent pursuit of urban sustainability. When ecologically friendly practices are paired with contemporary solid waste management procedures, the negative environmental effects of urbanization can be significantly reduced, public health can be enhanced, and natural resources can be preserved. We've looked at the various solid waste treatment techniques and how they could be applied to address issues with urban sustainability. It is evident that the best approaches for managing solid waste have been incorporated, sustainable, and context-responsive ones. Reusing resources decomposition and trash-to-energy conversion are two instances of circular economy initiatives that can help reduce the quantity of trash disposed of in landfills and mitigate the harmful impacts of waste disposal on the environment. Additionally, by leveraging state-of-the-art technology including anaerobic digestion and bioreactors, waste treatment methods can be made far more efficient. With the goal to more successfully tackle the issues related to urban sustainability, future research should focus on developing waste material treatment methods that are both more affordable and environmentally benign. This could entail looking for new uses for already-developed technologies, including combining waste into energy conversion with other environmentally friendly processes like biogas generation or using treatment of wastewater plant discharge as a nutrient-rich fertilizer. Moreover, in order to encourage public knowledge and participation in environmentally friendly waste disposal techniques, stakeholders and legislators should give priority to outreach and education programs. A legislative and policy environment that is supportive is essential for the achievement of any kind of solid waste treatment system. Clear rules, promotions, and standards for environmentally friendly waste management techniques must be established by governments and regulatory agencies. This could entail putting extended producer responsibility (EPR) laws into place to incentivize producers to create more environmentally friendly goods and packaging, or it could entail establishing goals for recycling and waste reduction. Governments can encourage investment and innovation regarding environmentally friendly solid waste treatment techniques by establishing legislative environments that are favorable to these activities. In order to achieve urban sustainability, modern waste treatment technologies must be adopted. By adopting environmentally friendly procedures, embracing the ideas of the "circular economy," and promoting public education and engagement, cities may decrease their impact on the environment, protect natural resources, and enhance public health. As the global population proceeds to move into cities, policymakers, stakeholders, and researchers must place a high premium on the development of novel and ecologically sustainable techniques for managing urban solid waste. There are various solid waste types that are harmful to the environment. Through the contamination of soil, water, air, and other biotic components, they lead to environmental pollution. Solid wastes come in different forms: industrial wastes (such as sewage sludge, dust, and various solid waste products), native solid wastes, and agricultural wastes (such as agricultural residues, animal dung, fruit peels, and wastes from thinning and pruning). In the present era, their management is a major concern. Solid waste management is accomplished through a variety of techniques, including open burning, land filling, and dumping into the sea. However, these management techniques are hazardous to the environment and are not eco-friendly. As science and technology advanced, scientists came up with a number of sustainable solutions to address issues related to managing solid waste and the global energy crisis. They developed the processes for producing fuel from animal dung, sewage sludge, and crop leftovers. They used techniques like burning, digestion, methanation, pyrolysis, and plasma gasification, among others, to turn solid waste into biogas, biodiesel, and electricity. These approaches are economical and thoroughly researched. Future research is being conducted to evaluate the more advanced technique of recovering energy from wastes with lower emissions in order to completely eliminate the problems related to climate change that are caused by the continued consumption of fossil fuels.

References

1. Kumar A, Samadder SR, Kumar N, Singh C (2018) Estimation of the generation rate of different types of plastic wastes and possible revenue recovery from informal recycling. Waste Management (New York, N.Y.), 79:781–790
2. Guerrero M, Re M, Kazimierski LD, Menéndez ÁN, Ugarelli R (2013) Effect of climate change on navigation channel dredging of the Parana River. Int J River Basin Manag 11(4):439–448
3. De S, Debnath B (2016) Prevalence of health hazards associated with solid waste disposal-A case study of kolkata, India. Procedia Environ Sci 35:201–208
4. Nowicki MA, Castro NJ, Plesniak MW, Zhang LG (2016) 3D printing of novel osteochondral scaffolds with graded microstructure. Nanotechnology 27(41):414001
5. Biedenbach DJ, Kazmierczak K, Bouchillon SK, Sahm DF, Bradford PA (2015) In vitro activity of aztreonam-avibactam against a global collection of Gram-negative pathogens from 2012 and 2013. Antimicrob Agents Chemother 59(7):4239–4248
6. Polyakova V, Degaev E, El Haddad P (2018) Reduction of ecological and economic risks in utilization of solid domestic wastes and construction waste. In: Paper presented at the MATEC Web of Conferences
7. Jiya A, Ijah UJ, Galadima M, Akpan U (2019) Isolation of microorganisms associated with biodegradation of household domestic wastes for biogas production in Niger State, Nigeria. J BioMedical Res Clin Pract 2(1):70–76
8. Quintas PY, Dotto GL, Da Silva SM, Escudero LB (2019) Removal of thallium from environmental samples using a raw and chemically modified biosorbent derived from domestic wastes. Environ Sci Pollut Res Int 26(31):32285–32297

9. Pickin J (2008) Representations of environmental concerns in cost–benefit analyses of solid waste recycling. Resour Conserv Recycl 53(1):79–85
10. Tanskanen P (2013) Management and recycling of electronic waste. Acta Mater 61(3):1001–1011

Assessing the Market for Recycled Plastics: A Predictive Modeling Approach

Daitri Tiwary

Abstract In this study, recycled plastics, specifically in the United States (US), has been studied to understand its growth as a market. Factors contributing to this growth include increased plastic consumption for lightweight product manufacturing, government initiatives to reduce plastic waste, and a growing awareness of environmental concerns. To gather data, an analysis of official websites from American companies dealing in various types of recycled plastics was conducted. The analysis was expanded by utilizing published data. The findings, derived from market share statistics, a value chain assessment, and an evaluation of suppliers, indicate an increasing reliance of industry companies on multiple facets of the recycling process. It seems as though vertical integration is emerging as a prevailing trend. Based on resin type, end-use application, source, and region, the recycled plastics market is segmented. The market is led by the packaging sector, securing a dominant 36% share. Up next are the construction and building sector at 20%, the electrical and electronics sector at 15%, the automotive sector at 12%, the textile sector at 10%, and the other sectors wrapping up at 7%. Industries such as construction and automotive experience a rise in plastic consumption for lightweight components, among the many factors that stimulate market growth. Moreover, there is an increasing trend towards environmental sustainability and the implementation of government incentives. Nevertheless, the market must overcome certain obstacles like inconsistent feedstock, price competition from virgin plastic, and the inadequacy of plastic product design. Technological advancements in the recycling sector highlight the significance of superior sorting and decontamination processes, including color sorting, AI-driven robotic sorting, and metal recovery using induction sorting.

Keywords Recycled plastics · Plastic consumption growth · Recycled plastic categories

D. Tiwary (✉)
International Finance, Birla Institute of Management Technology, 201306, Greater Noida, India
e-mail: daitri.tiwary@bimtech.ac.in

J. A. Parray et al. (eds.), *Sustainable Urban Environment and Waste Management*, Urban Sustainability, https://doi.org/10.1007/978-981-96-1140-9_7

1 Introduction

Hailing from the Greek word plastikos meaning "capable of being molded or shaped," the familiar term plastic describes a group of man-made or semi-synthetic substances made up of polymers. Able to be shaped through extrusion, molding, or compression into varied solid forms, plastics are often crafted from fossil fuel-based chemicals such as natural gas or petroleum. Lightweight, sturdy, versatile, and cost-effective, these substances have garnered extensive utilization and flexibility.

Each polymer chain is made of repeating units of monomers; these organic polymers are present in plastics and consist of carbon atom chains. The polymer's properties are decided based on the structure of these chains, which are made up of thousands of monomers.

Halogenated plastics, acrylics, polyesters, polyurethanes, and silicones are some of the classifications of plastics depending on their chemical structure. Chemical processes like cross-linking, condensation, and polyaddition are also methods used for categorization. Physical properties, such as density, glass transition temperature, tensile strength, hardness, and thermal resistance, are also important in classification. In addition, manufacturing and product design play a role in some plastic classifications This comprises conductive polymers, engineered plastics, biodegradable plastics, thermoplastics, thermosets, and elastomers. The chemical makeup of thermoplastics, such as polyethylene (PE), polypropylene (PP), polystyrene (PS), and polyvinyl chloride (PVC), changes when they are heated, allowing them to be repeatedly molded. Thermosets, or thermosettings, on the other hand, only melt and take shape once. They remain firm after they have solidified. Reheating them causes them to break down rather than melt because the thermosetting process causes an irreversible chemical reaction. Rubber is vulcanized as an illustration of this process.

The practice of gathering, classifying, and processing plastic waste to create new products is referred to as plastic recycling. It is a crucial strategy to limit the consumption of virgin plastic materials and lessen the environmental impact of plastic trash. Mechanical recycling and chemical recycling are the two primary categories into which plastic recycling procedures can be divided.

1.1 Mechanical Recycling

Mechanical recycling is the technique of processing plastic waste again into raw materials or products without changing the chemical structure of the material. Thermoplastics of almost all types can be recycled through this method without any major impact on the quality. It is the most common form of plastic recycling.

Mechanical Recycling of plastics can be divided into six stages beginning with waste collection and processing them through various stages to convert into pellets or flakes to be used by the manufacturers. (1) Collection: It is the initial stage of the recycling process for plastic. It entails gathering trash plastic from residences, establishments, and institutions. Either the municipal government or private businesses carry it out. Taking the waste plastic to local collection locations like designated recycling bins or facilities is another alternative for collection. It can be a bottle bank on a corner or a regional disposal facility for municipal solid trash that is both recyclable and non-recyclable. It is a crucial step because the more plastic that is collected, the more material will be available for recycling. (2) Sorting: Since there are several types of plastics, sorting is necessary to separate them based on attributes like color, thickness, or usage. Sorting improves factory productivity and prevents end-product contamination. (3) Washing: This phase of recycling is essential. It eliminates contaminants like labels, adhesives, dirt, and food residue that could damage recycled plastics or have an adverse effect on the recycling process. (4) Shredding: The plastics are fed into shredders after being washed to reduce their size so that they can be processed further for reuse. Any leftover contaminants can also be discovered and eliminated through shredding. Plastics are broken down into tiny bits during shredding, making handling them easier. (5) Identification: The class and quality of the plastic components are evaluated at this step. By floating the plastic pieces in a water container, they are first sorted according to their density. They are then subjected to an examination known as "air classification" in order to ascertain their thickness. The plastic components are put in a wind tunnel to accomplish this. The larger or thicker bits stay at the bottom, while the smaller or thinner pieces float into the air. (6) Compounding: It represents the last stage of the recycling process. It entails melting and crushing the shredded or downsized plastics to create pellets that may be used by manufacturers as finished goods.

1.2 Chemical Recycling

It consists of a group of recycling methods, including depolymerization, hydrocracking, gasification, and pyrolysis. Plastic trash undergoes chemical restructuring during chemical recycling. In this procedure, chemical, thermal, or catalytic methods are used to break down the long-chain hydrocarbons in the plastics into shorter hydrocarbons. The resulting shorter hydrocarbon molecules serve as a feedstock for the creation of new recycled polymers. Mechanical recycling is the most often utilized method because it is less labor-intensive, even though chemical recycling is more efficient at breaking down plastics into their smaller constituent molecules.

2 Literature Review

An attempt is made in this chapter to review the relevant literature which could serve as a necessary backdrop for the present study.

A study by Grand View Research [2–5] provides valuable insights into the current status and future prospects of substituting virgin plastics with recycled plastics. The global recycled plastics market was valued at USD 47.60 billion in 2022, with an expected compound annual growth rate (CAGR) of 4.9% from 2023 to 2030. The increasing consumption of plastics in lightweight components across various industries, including building and construction, automotive, electrical and electronics, and others, is driving the demand for recycled plastics. Polyethylene is the leading product segment in the recycled plastics market, accounting for more than 26% of the global revenue in 2022. The study suggests that plastic bottles will be the main source of raw material for recycled plastics because they are widely used to package carbonated beverages, water, oils, and medications. The study also suggests that the packaging sector will hold the maximum market share in terms of the application of recycled plastics in 2022. Polaris Market Research [6] also provides an overview of the global recycled plastic market. As per their study, the recycled plastics market was valued at USD 46.08 billion in 2021 and it is expected to grow at a CAGR of 4.7% in the future. The report also states that a number of nations around the world are bringing legislation to encourage the use of recycled plastics. According to the report, key competitors are concentrating on product innovation and partnerships to strengthen their positions in the fiercely competitive market.

A report by Allied Market Research [1] provides insights on the virgin resin market's growth potential and factors that affect the opportunities and trends of the sector, market in terms of value and volume, keeping in account the type of resin used and their application. The study also provides insights on the competitive environment and the key players involved in the sector. As per the study, the global virgin resin market is valued at $375.7 billion in 2021 and will grow at a CAGR of 3.5% from 2022 to 2031 and will reach $528.9 billion. The primary reason driving this growth will be investment in infrastructure, regulatory requirements, and awareness about hygiene. A Statista report from 2023 sheds light on the world's manufacturing of plastics. The estimate shows that, with 390.7 million metric tons of plastic produced in 2021, the world's plastic output has increased significantly over time. According to the report, plastics have transformed numerous industries due to their adaptability and affordability in a wide range of uses, particularly in the packaging industry. The study also shows that in terms of global plastics production, Asia has the highest proportion, followed by North America. Insights on the recycling and reuse of plastic waste in building materials including bricks, tiles, and blocks are provided by a study by Springer Link [7]. According to the study, using plastic waste that has been previously recycled offers a great prospect for sustainable growth. To fully exploit the potential of using plastic waste in construction materials, additional study is required to solve issues including standardization, regulatory approval, and widespread application.

3 Research Methodology

3.1 Data

This study is concentrated on the recycled plastics industry operating in the US between 2017 and 2023. This timeframe was selected because of the availability of data and capturing the latest scenarios of the market. The data has been obtained from the official websites of the plastic recycling companies which have been profiled for the purpose of the study. Data from World Trade Organization has also been taken to analyze the trade pattern and the dependence of the US on China for plastic recycling.

3.2 Methods

The methods adopted in this study have been divided into three sections; Supplier Landscape Assessment, Value Chain Analysis, and Market Overview. The details of aspects covered in each section are given below.

Supplier Landscape Assessment: In this section, the market structure i.e., whether the market is fragmented or consolidated has been assessed. For this, 15 key suppliers have been profiled covering parameters such as company name, company website, product portfolio, type of supplier (diversified or specialized), and geographic presence.

Value Chain Analysis: In this section, the value chain of the overall recycled plastics industry has been analyzed indicating companies operating across different stages of the value chain. Different stages of plastics recycling such as collecting, sorting, washing, resizing, separating, and compounding have been assessed to gauge if the companies undertake end-to-end processing, including grinding and pulverizing, or are specialized in specific activities. Supply sources of plastic scrap such as window frames, furniture, packaging, containers, floors, cable insulation, and pipes used for recycling have also been evaluated.

Market Assessment: In this section, the market size and growth outlook for the period of 2022–2027 of the US recycled plastics market have been assessed. Key market drivers—such as rise in the use of plastics for the producing lightweight components by various end-use industries, including construction, automotive, electrical, and electronics—and key challenges such as supply shortage faced by the end users have been assessed Market maturity i.e., whether the market is nascent, growing, or matured, and characteristics such as rising M&As resulting in industry consolidation have been evaluated. Aspects such as the correlation of recycled plastics with virgin plastics, and trade analysis of the US for virgin and recycled plastics have been covered. Key market and technology trends impacting plastic recycling have been assessed including the technologies being used for waste sorting and decontamination, such as separation/sorting equipment based on sensors, advanced infrared, eddy current, magnetism, etc. On the market side, capacity expansions and strategic partnerships with material recycling companies are some of the prominent developments which have been included in the study.

4 Analysis and Findings

4.1 Supplier Landscape Assessment

The recycled plastics market in the US is highly fragmented and decentralized as there are a large number of small and medium-sized players competing in the market. All the suppliers are specialized as they are only recycling plastic scraps and not manufacturing any products using recycled plastics. Most of the suppliers are operating in the US and some in Canada and Mexico as well.

The Tables 1 and 2 above shows the product portfolio of the key suppliers who have been profiled. Most of the suppliers are offering nearly all types of recycled resins. There are some suppliers who are focusing only on recycling one or two types of resins.

The companies are recycling both industrial as well as household plastic waste (Table 3).

4.2 Value Chain Analysis

There are six stages in the plastic recycling value chain. The figure below shows the flow diagram of plastic recycling (Fig. 1).

Various sources, including homes, businesses, and recycling facilities, contribute plastic debris for collection. Recycling programs at the curb, drop-off locations, or specialist collection services can all be used to collect it.

The plastic scrap then undergoes a sorting process where different types of plastic are separated based on their resin type or other identification methods. Sorting can be done manually or using automated technologies like optical sorting machines.

Once sorted, the plastic scrap is cleaned to remove contaminants like dirt, labels, and adhesives. Then the cleaned plastic scrap goes through shredding, and grinding to break it down into smaller pieces, facilitating further processing.

The cleaned and pre-processed plastic pieces are melted down using heat. The molten plastic is then forced through an extruder, which shapes it into thin strands or pellets. This step helps prepare the plastic for subsequent manufacturing processes.

The melted plastic can be used as raw material in various manufacturing processes. It can be molded into new products like bottles, containers, and packaging materials, or extruded into films, fibers, or pipes. To acquire the desired qualities, recycled plastic can also be mixed with virgin plastic. Either different companies or the same

Table 1 Landscape of key suppliers

Key suppliers of recycled plastics								
Sl. No	Company	Type of supplier	Industries catered	Recycling capacity (tonnes/year)	Annual production (tonnes/year)	Geographical presence	Based In	Revenue $ Million[1]
1	77 recycling	Specialized	NA	NA	NA	US	Massachusetts	NA
2	Nu-Pro Polymers, Inc.	Specialized	– Construction	NA	NA	US	Illinois	<5
3	Nu-Tech Polymers Co. Inc.	Specialized	NA	NA	NA	US	Ohio	<5
4	Akshar Plastic Inc.	Specialized	– Engineering – Commodity	NA	NA	US	Illinois	8
5	Patrick J. Kelly Drums	Specialized	NA	NA	1,814	US	Pennsylvania	23.6
6	Peninsula Plastics Recycling, Inc.	Specialized	NA	NA	30,000	US	California	18.5
7	Prime Plastic Products, Inc.	Specialized	NA	NA	30,000	US	California	<5
8	Resource Plastics Inc.	Specialized	NA	NA	34,019	US	Illinois	10.2
9	True Plastics Recycling	Specialized	– Films – Bags	21,768	NA	US	Texas	<5
10	West Michigan Compounding LLC	Specialized	– Automotive	45,359	NA	US	Michigan	12.1
11	Talco Plastics, Inc.	Specialized	NA	NA	NA	US	California	11.8

(continued)

Table 1 (continued)

Key suppliers of recycled plastics

Sl. No	Company	Type of supplier	Industries catered	Recycling capacity (tonnes/year)	Annual production (tonnes/year)	Geographical presence	Based In	Revenue $ Million[1]
12	KW Plastics Recycling	Specialized	– Personal Care – Automotive – Agriculture	294,835	NA	US	Alabama	116
13	B&B Plastics Inc.	Specialized	– Injection Molders – Film/Sheet Extruders Thermoformers	NA	NA	– US – Mexico – Canada	California	6.8
14	ACI Plastics, Inc.	Specialized	NA	NA	NA	US	Michigan	7
15	Mega Recycling and Compounding Services LLC	Specialized		NA	NA	US	Illinois	<5

The list is non-exhaustive; *NA* Not Available. The geographic presence is based on the recycling plant location

Table 2 Products offered by key suppliers

Company wise details													
Company	77 Recycling	Nu-Tech Polymers Co. Inc.	Akshar Plastic Inc.	Patrick J. Kelly Drums	Peninsula Plastics Recycling	Prime Plastic Products, Inc.	Resource Plastics Inc.	True Plastics Recycling	West Michigan Compounding LLC	Talco Plastics, Inc.	KW Plastics Recycling	B&B Plastics Inc.	ACI Plastics, Inc.	Mega Recycling and Compounding
Products	– PC – ABS – PP – PS – PA – PMMA – POM – PVC – PBT – LDPE – HDPE	– Prime Resins – Resins – Custom Formulated Resins – Regrind Resins	– ABS – HIPS – PETG – PVC – LDPE – LLDPE – HDPE – PET – PP – XLDPE – NYLON 6 – PVS – PS – EPOXY	– Steel Drums – Poly Drums – Fibre Drums	– PET	– LDPE – HDPE – NYLON 6&6, 6 – ABS – PET – PP	– ABS – PS – PET – PETG – PVC – PE – PP – TPO – PC – Acrylic	– LDPE – LLDPE	– ABS – ASA – NYLON – PBT – PC – PP – TPE	– PE – PP – PS – ABS – PVC – Acrylic – PC – Noryl – Acetals – NYLON – Santoprene	– PE – PP	– HDPE – LDPE – HIPS – GPPS – CO – PP – PET – PVC Engineering Grade: – PC – PC/ABS	– LDPE – ABS – NYLON: 6&6, 6 – PC – HDPE – GTX – PTE	– PP – HDPE – LDPE – NYLON: 6&6, 6 – PBT – Acetal – PC – ABS – TPE – TPS – TPU – Noryl – K-Resin

Table 3 Types of plastic scrap recycled by the suppliers

Company wise details														
	77 Recycling	Nu-Tech Polymers Co. Inc.	Akshar Plastic Inc.	Patrick J. Kelly Drums	Peninsula Plastics Recycling	Prime Plastic Products, Inc.	Resource Plastics Inc.	True Plastics Recycling	West Michigan Compounding LLC	Talco Plastics, Inc.	KW Plastics Recycling	B&B Plastics Inc.	ACI Plastics, Inc.	Mega Recycling and Compounding
Source of raw material	– Clean post consumer scrap – Lawn Toys – Patio Furniture – Garden Hoses – Pool Liners	NA	ABS: –Taillights – Automobile Hub Caps – Wheel Covers PC: – CDs/DVDs – Bulk Water Bottles – Baby Bottles LDPE: – Bread Bags – Squeezable Bottles	– Poly Drums – Totes	– PET Products	HDPE (Film/ Scrap): – Grocery Bags – Store Bags – Mixed Film Printed, Non Printed HDPE: – Buckets – Totes – Pallets – Crates – Bread Trays – Bottles	Post Industrial Plastics: – ABS – PS – PET – PETG – PVC – PE – PP – TPO – PC – Acrylic	– Films – Bags – Startup Rings – Purge – Bricks – Foam	– Post Industrial Plastic Scrap	– Engineering and Commodity Resins in all forms – Regrind Lumps – Chunks – Bales – Pellets – Sheets	– HDPE Natural Bales – HDPE Pigmented Bales – HDPE Bulky Rigid Bales – PP Tubes – PP Lids – PP Bottle Bales – PP Battery Chips	– HDPE – LDPE – HIPS – GPPS – CO – PP – PET – PVC Engineering Grade: – PC – PC/ABS	– PP – PA – PC – HDPE – ABS – PMMA – PBT – TPE	– PP – HDPE – LDPE – NYLON: 6&6, 6 – PBT – Acetal – PC – ABS – TPE – TPS – TPU – Noryl – K-Resin

Fig. 1 Various stages of the plastic recycling process

company can carry out each of the steps. The suppliers involved in the various phases of the recycling process are listed in the Table 4.

The analysis shows that most of the companies are carrying out the various stages of the recycling process by themselves. Some of the companies are even involved in the plastic waste collection process. This indicates that the companies are vertically integrated.

Table 4 Recycling activities undertaken by the suppliers

Sr. No	Company name	Activities undertaken by the suppliers				
		Collection	Sorting	Washing	Shredding	Compounding
1	77 Recycling	NA	NA	NA	✓	✓
2	Nu-Pro Polymers, Inc	✗		✓	✓	✓
3	Nu-Tech Polymers Co. Inc	✓	✓	✓	✓	✓
4	Akshar Plastic Inc	✗	✓	✓	✓	✓
5	Patrick J. Kelly Drums	✗	✓	✓	✓	
6	Peninsula Plastics Recycling, Inc	✗	✓	✓	✓	✓
7	Prime Plastic Products, Inc	✓	✓	✓	✓	✓
8	Resource Plastics Inc	✗	✓	✓	✓	✓
9	True Plastics Recycling	NA	NA	✓	✓	✓
10	West Michigan Compounding LLC	✓	✓	✓	✓	✓
11	Talco Plastics, Inc	✓	✓	✓	✓	✓
12	KW Plastics Recycling	✓	✓	✓	✓	✓
13	B&B Plastics Inc	✗	✓	✓	✓	✓
14	ACI Plastics, Inc	✓	✓	✓	✓	✓
15	Mega Recycling and Compounding Services LLC	✓	NA	NA	✓	✓

NA Not Available

4.3 Market Assessment

Overview: The total plastic waste produced globally in 2021 stood at approximately 400 million metric tons. Out of this, the US alone contributed nearly 40 million metric tons which is almost 10% of the global plastic waste generated (Fig. 2).

Out of the total plastic waste produced in the US, only 5–6% gets recycled. This is because of the inconsistent collection and sorting processes, plastic waste contamination, and poor product design (Fig. 3).

The US recycled plastic market stood at $ 3.31 billion in 2021 and is expected to grow at a CAGR of 4.1% between 2022 to 2027. This growth rate of the recycled plastics market is expected to be propelled by the increasing plastic consumption for the production of lightweight products coupled with increasing environmental concerns and government efforts aimed at reducing plastic waste (Fig. 4).

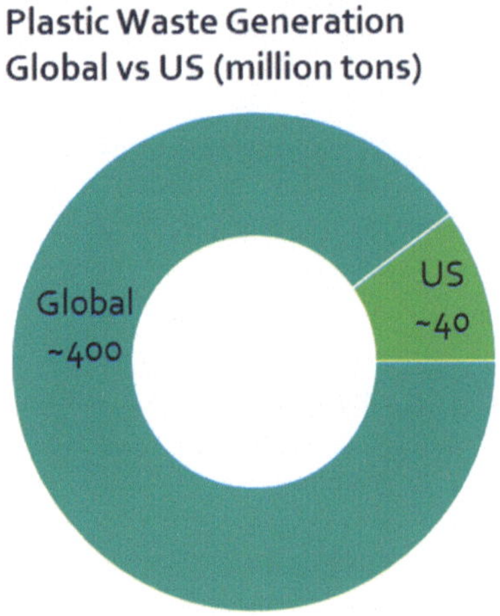

Fig. 2 Share of plastic waste produced by US

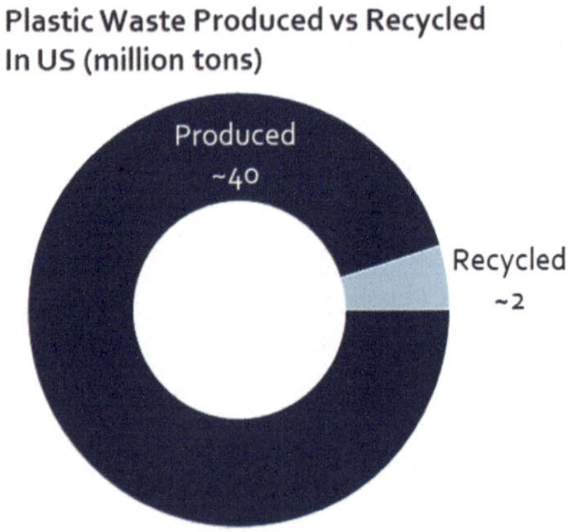

Fig. 3 Proportion of plastic waste recycled in the US out of the total plastic waste produced

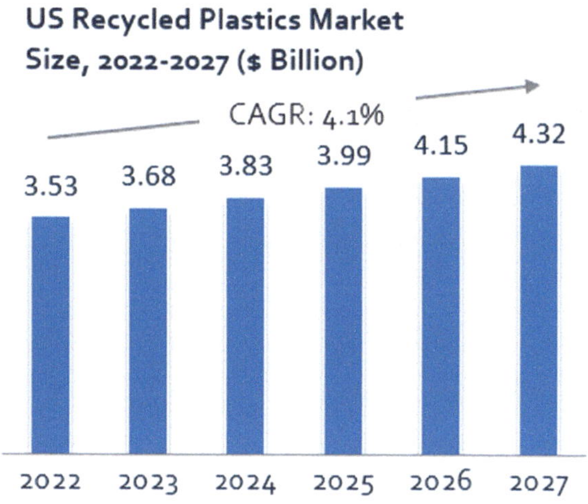

Fig. 4 Expected growth rate and market size of US recycled plastic market, 2022–27

4.4 *Trade Analysis*

For recycling the plastic waste produced, the US has always depended on other nations. It outsourced its waste to Asian countries. The majority of waste paper and plastic had been imported from China. In 2017, the US and China imported scrap plastic from each other to the tune of $11.18 million. China established a number of environmental protection regulations between 2013 and 2018, including the Blue-Sky Policy, National Sword Policy, and Green Fence Policy, to prevent the unlawful recycling and smuggling of recyclables and safeguard the environment. In addition to the 6.5% Most Favored Nation duty on plastic waste imported from any WTO member, China placed a 25% tariff on US plastic and paper waste in 2017 as a result of the trade tensions between the US and China. As a result, from 2017 to 2021, the US-China trade in plastic scrap drastically decreased (Figs. 5 and 6).

Virgin and Recycled Plastics Market in the US, 2021: The US plastic and resin manufacturing market size stood at $74 billion in 2021. The recycled plastic market

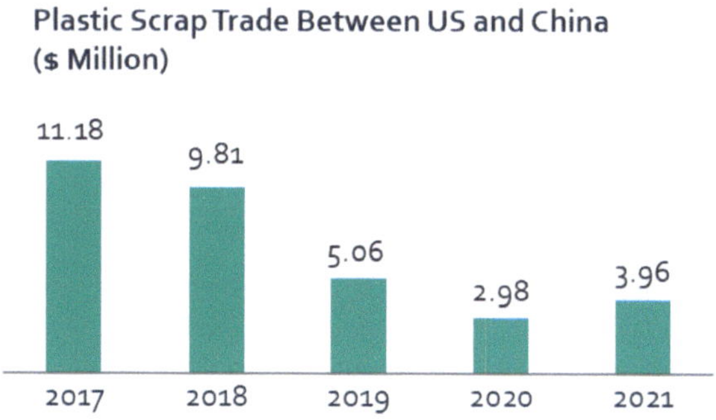

Fig. 5 Decline in the plastic scrap trade between US and China, 2017–21

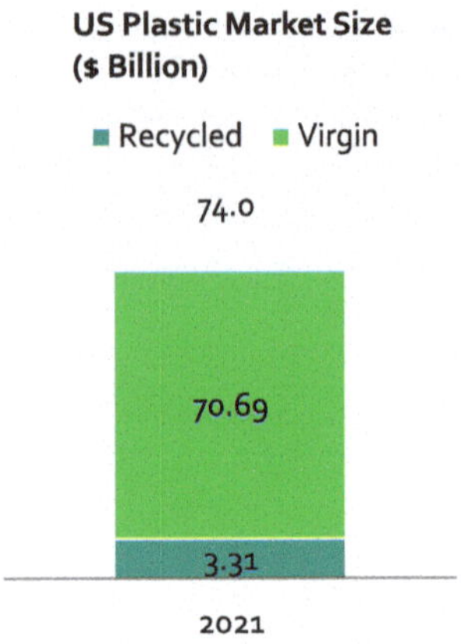

Fig. 6 Market share of virgin and recycled plastics in the US, 2021

contributed \$3.31 billion whereas virgin plastic contributed \$70.69 billion. The recycled plastic market is a growing one and is expected to grow at a CAGR of 4.1%, it is still in a nascent stage when compared to the virgin plastic market which makes up approximately 96% of the total plastic market.

4.5 *Market Segmentation*

The recycled plastics market is classified into four categories; based on resin type, source, end-use, and region (Figs. 7 and 8).

Market Segmentation Based on End-Use Application:

With a 36% utilization market share, the packaging sector rules the recycled plastic market. Recycled plastic is used to make food and beverage bottles, and non-food containers. Building and Construction sector follows with a 20% market share. It

Fig. 7 Market segmentation of recycled plastics

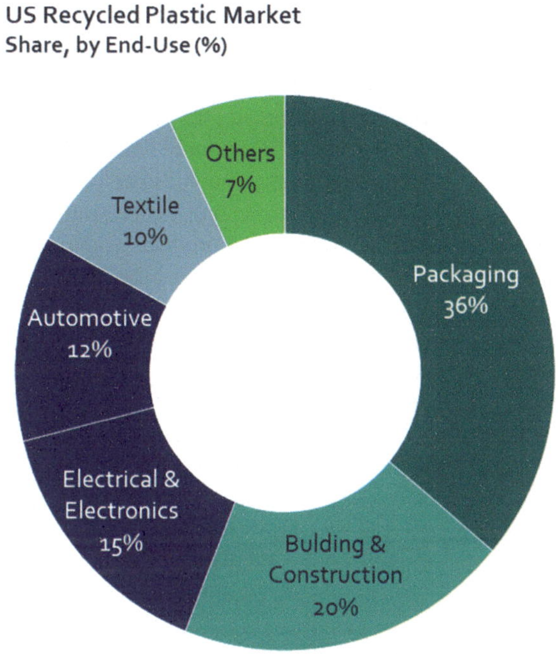

Fig. 8 Sector-wise market share of recycled plastics

is used in construction brick making, indoor insulation, sidewalks and driveways, PVC windows, fences, floor tiles, etc. The Electrical and Electronics, and Automotive sector holds 15 and 12% respectively. Recycled plastic is used in cable and wire insulation, electronic equipment such as telephone handsets, keyboards, and computer housings, and in the automotive sector, in wheel liners, splash guards, air deflectors, etc. The textile sector holds approximately 10% of the market share. Recycled plastic is used to make fabric for t-shirts, leggings, etc. The remaining 7% is comprised of other sectors which include agriculture, healthcare, etc. Face masks, IV bags, catheters, etc. are made using recycled plastics.

5 Conclusion

5.1 Government Incentives

Various state governments provide tax incentives in the form of tax credits and tax exemptions to boost the recycling business.

Tax Incentives provided by various State Governments for recycling businesses: The State of Arizona provides income tax credit equalling 10% of the installed cost of equipment to individuals and corporates processing post-consumer recyclables, the amount of credit claimed in a tax year cannot exceed the lesser of $5,000 or 25% of the current year's tax liability without applying for the credit. The State of Delaware provides a recycling investment income tax credit of $500 for each $100,000 invested and $500 for each qualified employee added, the aggregate amount of credits claimed in any given year may not exceed fifty percent of the firm's tax liability. The Recycling Equipment Tax Credit Program of Arkansas, encourages recycling by allowing Arkansas taxpayers a 30% state income tax credit for the purchase of equipment used exclusively for collecting, separating, processing, modifying, converting, or treating solid waste or for manufacturing products containing at least 50% recovered materials of which at least 10% is post-consumer waste. The State of Louisiana provides a tax credit to any taxpayer who purchases qualified new recycling manufacturing or process equipment and/or qualified service contracts and has been approved by the Department of Environmental Quality. The State of Utah provides a 20% income tax credit (up to $2,000) on eligible operating expenses, including payments made to third parties for rent, wages, supplies, tools, test inventory, and utilities, for establishing and operating recycling or composting technology in Utah. The State of South Carolina provides sales tax exemptions for Machinery and equipment, and applicable repair parts, used in recycling. The State of Virginia provides a tax credit equal to 20% of the original total capitalized cost of the purchase price for machinery and equipment for processing recyclable materials, the total credit allowed cannot exceed 40% of the Virginia income tax liability in any taxable year.

Similarly, other state governments also provide tax incentives to individuals as well as businesses to promote plastic waste recycling. This acts as a key driver in the recycled plastics market.

5.2 *Market Drivers and Challenges*

5.2.1 Market Drivers

(1) Increasing Plastic Consumption for Lightweight Components: (1) Construction Industry: PET plastic waste is mixed with foundry sand in the ratio 20:80, 30:70, and 40:60 to get lightweight bricks with enhanced tensile and compressive strength. (2) Automotive Industry: The use of plastic components in automobiles makes them more lightweight and enhances fuel efficiency. In June 2018, Ford announced that each of its vehicles contains 250 bottles of recycled plastics which is equivalent to the use of 1.2 billion bottles per year. Renault Group is engaged in the use of recycled materials. Megane II; a model launched by Renault Group in 2014 has 30% of its components made from recycled plastics.

(2) Growing Environmental Awareness and Sustainability Initiatives: Growing awareness among consumers, businesses, and governments about the environmental impact of plastic waste and the need for sustainable solutions has led to increased demand for recycled plastics as a way to reduce plastic waste, conserve natural resources, and lower the carbon emissions.

(3) Government Incentives: Governments and regulatory bodies are implementing favourable policies and incentives to promote plastic recycling. Reclaimed plastic tax credit of 50% of the plastic recycling capital investment taken at a rate of 10% per year for 5 years. Pollution control facility tax credit of 50% of the recycling equipment and facility capital cost taken at a rate of 5% per year for 10 years.

5.2.2 Market Challenges

(1) Insufficient and inconsistent feedstock: Lack of availability of sufficient and good quality feedstock due to inconsistent collection and sorting processes, contamination of plastic waste, and lack of infrastructure for recycling results in supply shortages and variability in product quality
(2) Price competition from virgin plastic: Low-cost virgin plastic is a big challenge for the recycled plastics market because of the economies of scale achieved in the production of virgin plastics. Also, the long procedure of recycling and handling material between different stages of the recycling process increases the cost of recycled plastics. High contamination level in the recovered plastic feeds increases the cost of sorting and processing and thus reduces its ability to compete with the cheap virgin resin.
(3) Improper design of current plastic products: Many products are designed in such a way that it becomes difficult to separate and recycle the plastic. Different types of plastics combined together, glues and adhesives are applied to the plastic. Such product designs impact the management of plastic throughout its lifecycle and make them unfit for recycling.

5.3 Technology and Market Trends

Technology Trends: Advanced sorting and decontamination technologies are being incorporated by recycling companies to enhance the process efficiency and minimize human dependence. Some of the sorting and decontamination technology innovations in waste handling and recycling operations are given below: (1) Colour Sorting: It is also known as Optical Sorting, Digital Sorting, or Electronic Colour Sorting. It is used to distinguish between coloured and colourless PET and HDPE flakes and separate them by colour. (2) Robotic Sorting: It uses AI and Robotics to sort the plastics. It uses advanced cameras and high-tech computers to detect different objects and place them in their respective bins. (3) Induction Sorting: It is an addition to magnetic, and eddy-current separation for recovering residual metals from a mix of materials. Metal detectors are arranged over the conveyor belt; the inductance changes in the magnetic coil when metal passes by it is blown away using compressed air. (4) Near Infra-Red Sorting: It is the recycling industry's preferred plastic sorting technology. It accurately identifies the different types of polymers that are in use today and separates them. (5) Chemical Recycling: It comprises a set of technologies (pyrolysis, gasification, hydro-cracking, de-polymerization). In this method, the chemical structure of plastic waste is changed. Long-chain hydrocarbons are broken into shorter ones to be used as feedstocks to produce recycled plastics. (7) Dissolution Recycling: In this method, the polymer present in mixed plastic waste is selectively dissolved in a solvent to separate it from the waste without changing its chemical nature. This technique is used in recycling PVC, PP, PS from mixed, multi-material waste.

Recycling companies are adopting these innovative and advanced sorting technologies to improve their efficiency. Near Infra-Red sorting technology is the preferred one because of its ability to identify various types of polymers that are used today.

There has been an upward tick in the market trend with major players entering the recycling industry. KW Plastics Recycling, in 2021, started a polypropylene wash line to increase its processing capacity by 45,360 tons/year with an investment of $18.75 million owing to the increasing demand. West Michigan Compounding, LLC., in 2018, added over 41,000 sq. ft. to its production floor along with 3,600 sq. ft. to the front office to boost its capacity to meet the increased demand. Peninsula Plastics Recycling, Inc., in 2022, expanded its PET recycling by 50% with the help of a loan of $ 4.5 million received from the State of California. PureCycle Technologies, on March 16, 2023, announced a partnership with Formerra; an engineering materials distributor; as an authorized North American distributor of PureCycle's Ultra Pure Recycled (UPR) PP resin. ACI Plastics, Inc., in February 2023, started its new recycling facility in Flint, Michigan that has the capacity to process 10.9 tons/year of post-consumer plastic film. Mega Recycling and Compounding Services LLC, in 2019, expanded its processing capabilities by adding three single-screw extruders to its operations.

Owing to the increasing demand for recycled plastics in various applications companies have been primarily resorting to organic growth options i.e., going for capacity expansion to capitalize on the available market opportunities. Some companies are also opting for strategic partnerships to expand their reach but capacity expansion is the common trend among the companies.

5.4 The Way Forward

The implementation of chemical recycling techniques such as dissolution, pyrolysis, gasification, and depolymerization offers the potential to fully or partially substitute virgin plastics with recycled plastics. These methods break down waste plastics into their smallest constituent parts, essentially in the form of oil or gas, from which the virgin plastics are made in the first place. This offers a substantial chance to lessen the need for virgin plastics, save natural resources, and lessen the negative effects of plastic waste on the environment. It is crucial to remember that these sophisticated recycling methods frequently call for substantial amounts of waste feedstock and can be difficult to put into practice on a modest scale. As a result, the collection and sorting procedures are put under even more strain. These procedures are crucial for ensuring a steady stream of recyclable plastic garbage. Additionally, the oil or gas created by the chemical recycling process would need to go through a fractional distillation process in order to produce various types of resins, necessitating a partnership between crude oil refining firms and the recycling companies. Enhancing the collecting and sorting infrastructure is essential to maximizing the advantages of chemical recycling and raising the amount of plastics recycled. This entails enhancing waste management systems, putting in place effective collection programs, and educating consumers about the significance of appropriate plastic disposal.

To maximize total recycling rates, integrating chemical recycling technology with already-in-use recycling techniques can be a smart move. By combining mechanical recycling, which involves traditional sorting and processing techniques, with chemical recycling, organizations can optimize the utilization of plastic waste and achieve higher recycling efficiencies.

References

1. Allied Market Research (2022) Virgin resin market. Retrieved from https://www.alliedmarketresearch.com/virgin-resin-market-A31754
2. EPA (2023) Recycling market development. Retrieved from https://archive.epa.gov/wastes/conserve/tools/rmd/web/html/rec-tax.html#sc
3. Grand View Research (2023) North America plastic recycling market report. Retrieved from https://www.grandviewresearch.com/industry-analysis/north-america-plastic-recycling-market-report#:~:text=b.,USD%203%2C273.2%20million%20in%202021.
4. Grand View Research (2023) Recycled plastics market. Retrieved from https://www.grandviewresearch.com/industry-analysis/recycled-plastics-market
5. Grand View Research (2023) Recycled polyethylene terephthalate (PET) market. Retrieved from https://www.grandviewresearch.com/industry-analysis/recycled-polyethylene-terephthalate-pet-market
6. Polaris Market Research (2022) Recycled plastics market. Retrieved from https://www.polarismarketresearch.com/industry-analysis/recycled-plastics-market
7. Springer (2021) Title of the article. Journal Name, Volume(Issue), page range. DOI

Environmental Justice as a Practical and Sustainability Issue in Sub-Saharan Africa: Gaps and Options

Innocent Chirisa, Tinashe Magande, Ericcson Tinashe Mapfumo, and Archeford Munyavhi

Abstract In environmental discourses, environmental justice has grown to be a significant problem. This paper comes from the background that major environmental agreements like the United Nations Framework Convention on Climate Change () and the Kyoto Protocol, increasingly include concerns for environmental equality and justice. This study situates the questions of environmental justice within the larger context of environmental sustainability and the current discussions of justice theories. It examines the environmental justice movement that has recently gained popularity in the USA. With a strong social justice component, this specific grassroots movement has redefined the sustainability agenda in Sub-Saharan Africa. It has characteristics with social justice movements for environmental justice in developing countries sometimes known as "environmentalism of the poor." This paper critically evaluates several of the well-known conceptions of justice based on various principles of justice such as need, desert, and entitlement, using a critical discursive technique. current discussions on universalism against particularism, "abstract" liberalism versus communitarianism, and some other critical viewpoints on justice are

I. Chirisa (✉)
The Vice Chancellor's Office, Zimbabwe Ezekiel Guti University, Bindura, Zimbabwe
e-mail: chirisa.innocent@gmail.com

Department of Urban and Regional Planning, University of the Free State, Bloemfontein, South Africa

T. Magande
Department of Urban and Regional Planning, University Johannesburg, Johannesburg, South Africa

E. T. Mapfumo
School of Health and Human Science, University of Huddersfield, UK & Carnegie, Huddersfield, UK

School of Education, Leeds Beckett University, Leeds, UK

A. Munyavhi
Department of Agribusiness and Management, Faculty of Agribusiness and Entrepreneurship (FAE), Marondera University of Agricultural Sciences and Technology (MUAST), Marondera, Zimbabwe

J. A. Parray et al. (eds.), *Sustainable Urban Environment and Waste Management*, Urban Sustainability, https://doi.org/10.1007/978-981-96-1140-9_8

examined. The paper recommends the need for a more expansive definition of environmental justice that recognises individual differences in different countries, while also being aware of the global nature of environmental issues. This paper concludes that environmental justice has to consider differential effectsfrom disproportionate contributions to environmental damages. It contends that a theory of justice that takes this reality into account must likewise take distinct obligations into account.

Keywords Environmental equality · Africa · Environmental demands · Conflicts · Disputes · Society and natural resources

1 Introduction

The need for justice in society seems to be a constant.[1] It is frequently seen as the most coveted social virtue. Justice-related cries are heard everywhere showing that there are unsolved problems at many levels of social interaction. Conflicts and disputes reveal that society and its institutions must continue to address the issues that serve as the foundation and inspiration for these grievances and cries for justice.[2,3] These are the concerns that characterize and frequently support claims of justice made by persons who have experienced injustice as they understand it but sometimes not how others do. Therefore, justice is a historically persistent phenomenon that denotes the existence of damage, exploitation, and oppression in society on the one hand and a desire to address these shortcomings on the other.[4]

Equity and environmental justice should be the main long-term objectives of urban climate strategies.[5,6,7] This is so because equity promotes social capital, sustainable urban development, and human well-being, all of that improve a city's ability to adapt to climate change. It is crucial to have access to land in safe areas and to secure housing, essential services and infrastructure that lowers danger.[8,9] To avoid rigid and top-down solutions, cities must encourage and disseminate a science-based policy-making process that incorporates the interests of several stakeholders. Participatory approaches that integrate community members' opinions on the viability of resilience targets can achieve this.

---

[1] Schaefer [1].

[2] Botterhuis [2].

[3] Schaefer [1].

[4] Ibid.

[5] Nyamwanza [3].

[6] Menton [4].

[7] Isgren [5].

[8] Menton [4].

[9] Isgren [5].

2 Conceptual Framework

The conceptual framework for this study is based on the strand of the eco-justice concept that also includes ecological justice (concerning nature and particular habitats) and species justice (mainly focusing on nonhuman creatures), which is environmental justice (concerning people).[10] In terms of access to and use of certain natural resources in designated geographic regions, and the effects of certain social behaviours and environmental risks on particular populations, it refers to the equal distribution of habitats among people (as defined based on class, occupation, gender, age, tribe, caste and ethnicity). Regarding toxic surroundings, it is mainly worried about the health and welfare of people, groups, and communities.[11,12,13]

Actual incidents and occurrences that have a detrimental impact on people and their communities are what gave rise to environmental justice movements.[14] The motivation is tangible and based on experience. This study denotes that studying environmental justice requires participating in critical criminology that is informed by notions of social and ecological justice, to "green criminology," a discipline of academic inquiry.[15] The environmental justice framework that aims to stop environmental risks is based on several connected claims and concepts.[16] These principles place a strong emphasis on notions like social justice, which hold that everyone has a right to protection against environmental deterioration, and damage reduction (that focuses on eliminating a threat before harm occurs).[17] Measurements of factors like human exposure to environmental toxins and sociological analyses of damage and risk distributions across various population groups are necessary for each of these domains.

The paper makes use of the environmental justice framework that aims to stop environmental risks and is based on several connected claims and concepts.[18] These principles place a strong emphasis on notions like social justice that hold that everyone has a right to protection against environmental deterioration, and damage reduction (that focuses on eliminating a threat before harm occurs). Measurements of factors like human exposure to environmental toxins and sociological analyses of damage and risk distributions across various population groups are necessary for each of these domains. Additionally, to the environmental justice framework, "parties applying for operating permits for landfills, incinerators, smelters, refineries, chemical plants, and similar operations must prove that their operations are not harmful to human health,

[10] White [6].
[11] Bullard [7].
[12] Bullard [8].
[13] Pellow [9].
[14] Isgren [5].
[15] Ibid.
[16] Bullard [8].
[17] Pellow [9].
[18] Bullard [8].

will not disproportionately affect racial and ethnic minorities and other protected groups, and are nondiscriminatory".[19]

3 Literature Review

The idea of environmental justice first surfaced in the United States of America in the latter half of the twentieth century, founded in impoverished communities of colour's attempts to fight unequal exposure to risk from industrial pollution.[20,21] The concept of environmental justice is complex, anchored in sustainable development, and involves managing the world's resources for both the present and the future while taking into account the environment, the economy, and social justice.[22] Deconstructing the phrase, "environment" may be roughly described as the situations, things, and people around someone or something.[23]

This study defined 'Justice' as an ethical term that is defined as the standard of fairness in that like cases are treated equally, the equitable distribution of burdens and benefits between cases conceived as similar, or the idea that the severity of the punishment should be commensurate with the related transgression.[24] Thus, "injustice" might be defined as the absence of the justifications outlined by Patel.[25] There is broad consensus that the core concept of environmental justice is the equitable allocation of environmental externalities. Environmental injustice is the uneven distribution of the planet's resources, to Dominelli.[26] Environmental justice is distinguished from ecological justice by Low and Gleeson in that the former relates to the equitable distribution of habitats among people, whilst the latter deals with the interaction between humans and the rest of nature.[27] Environmental resources and features that are both advantageous and unfavourable are referred to as "environments" in discussions of environmental justice.[28]

Numerous times, the Brundtland Report asserts causal relationships between environmental deterioration, inequality, and poverty.[29] It does make an inadequate effort

---

[19] (ibid., 28–29).

[20] Patel [10].

[21] Rogge [11].

[22] Ibid.

[23] Dictionary, Penguin English. "The Penguin English Dictionary." (2007).

[24] Dictionary, Collins English. "Harper Collins." *London & Glasgow* (2000).

[25] Patel [10].

[26] Dominelli, Lena. "Environmental justice at the heart of social work practice: Greening the profession." *International Journal of Social Welfare* 22, no. 4 (2013): 431–439.

[27] Low, Nicholas, and Brendan Gleeson. "Situating justice in the environment: the case of BHP at the OK Tedi Copper Mine." *Antipode* 30, no. 3 (1998): 201–226.

[28] Patel [10].

[29] Brundtland, Gro Harlem. "Brundtland report. Our common future." *Comissão Mundial* 4, no. 1 (1987): 17–25.

to broaden the discussion of these topics, which are frequently left out of the mainstream environmental discourse and the many conservation groups. For instance, the concept of sustainable development as promoted by international groups like the International Institute for Environment and Development (IIED) and particularly by the IUCN's World Conservation Strategy (WCS), is distinctly limited and lacks some of the more social and political insights.

Although vital, the overemphasis on conservation and preservation obviates the need for a comprehensive assessment of the situation (Adams, 1990: 51 in O'Riordan 1993, p. 50) drew attention to this flaw and wrote that the WCS was "pious, liberal and benevolent, necessarily ideological and disastrously naive." Thus, environmental concerns are frequently viewed as singular problems and largely separated from their larger socioeconomic contexts. The assertions on the connections between poverty and environmental destruction not only change the subject but also exacerbate problems with justice. People who are hungry and underprivileged frequently ruin their local surroundings to live and it is pointless to try to deal with environmental problems without a larger vision that covers the reasons underpinning world poverty and worldwide inequality.[30] It is unclear how much and why poverty "pollutes" the environment. Who pollutes whose environment matters just as much, if not more, in the context of claims of environmental justice as this topic? These challenges have influenced the goals of several sustainable development movements with significant social justice components across the world, even though they are themselves reflective of different environmental understandings and associated social and political discourses. The "environmental justice" movement in the United States is comparable to a social justice movement in that it is not just focused on the preservation and conservation issues that dominate the agendas of so many other powerful Northern sustainability organisations.[31] The rising concern that the goals of social justice and sustainability may not be the same is rooted in the seeming lack of a significant conservation component in such justice groups.

Numerous studies on particular occurrences and victimisation patterns have shown that some persons are more susceptible than others to suffering negative effects from environmental issues.[32] This is clear from the placement of toxic waste dumps, the severity of air pollution, chemical mishaps, the availability of safe, clean drinking water, and other factors.[33,34,35] No matter where these environmental inequities

[30] Douglas [12].
[31] Agyeman [13].
[32] Ibid.
[33] Chunn [14].
[34] Saha [15].
[35] Williams [16].

occur—in the United States, Canada, India, or Australia—it is the poor and disadvantaged that suffer disproportionately from them.[36,37,38,39] These communities also bear the brunt of the negative effects of resource extraction. For instance, oil, lumber, and minerals are exploited in many locations across the world where minority or indigenous peoples reside in ways that harm local ecosystems and traditional cultures and livelihoods.[40,41,42]

When considering the anthropocentric causes of climate change, the sheer size of mining activities and how they are conducted are also of utmost importance.[43,44] The influence of mining on greenhouse gas emissions can be significant. By way of illustration, the Alberta Tar Sands project in Canada is infamous for its significant contribution to greenhouse gas emissions because of its massive size and open-cut mining techniques.[45] Recent legal actions in Australia against extensions to already-existing open-cut coal mining operations in Queensland have brought attention to the environmental damage, especially greenhouse gas emissions, caused by such extensions.[46]

A crucial point is that the hunt for more rare natural resources is pushing businesses into uncharted territories of mineral and gas exploration and technical exploitation.[47] There are various negative hazards associated with this. First, there are risks and dangers related to operations like deep-offshore drilling (as demonstrated by the collapse of the BP oil rig in the Gulf of Mexico); new extraction techniques come with new risks and possible consequences. Second, it is less probable that there will be effective control of mining operations the more distant and marginal the regions that are mined are, and the more dependent those populations are on mining for their local economies. In any scenario, concerns about the widespread creation of greenhouse gas emissions continue to be of considerable concern.

---

[36] Bullard [8].

[37] Rush [17].

[38] Engel [18].

[39] Walker, Brian, Lance Gunderson, Ann Kinzig, Carl Folke, Steve Carpenter, and Lisen Schultz. "A handful of heuristics and some propositions for understanding resilience in social-ecological systems." *Ecology and society* 11, no. 1 (2006).

[40] Brook [19].

[41] Gedicks, Frederick Mark. "Suitability Claims and Purchases of Unrecommended Securities: An Agency Theory of Broker-Dealer Liability." *Ariz. St. LJ* 37 (2005): 535.

[42] Schlosberg, David. *Defining environmental justice: Theories, movements, and nature*. OUP Oxford, 2007.

[43] Brook [19].

[44] Gedicks, Frederick Mark. "Suitability Claims and Purchases of Unrecommended Securities: An Agency Theory of Broker-Dealer Liability." *Ariz. St. LJ* 37 (2005): 535.

[45] Klare, Michael T. "Energy security." In *Security Studies*, pp. 557–574. Routledge, 2012.

[46] Millner [20].

[47] Agyeman [21].

Africa is most susceptible to climate change and the least prepared to adapt because of ongoing development issues.[48] Global climate solutions must comprehensively address Africa's unique vulnerabilities if success is to be realised.[49] Africa has made the least effort to combat climate change since it is the least industrialised continent. However, it is suffering the most from the effects of climate change on a worldwide scale. Events related to climate change such as temperature rise and irregular precipitation, have an impact on Africa similarly to other geographical areas. The ten nations that are most vulnerable to climate change worldwide are found in Africa, where these effects are felt most keenly. Due to several pre-existing issues including poverty, food insecurity, displacement, and violence, there is a pervasive vulnerability that feeds a vicious cycle where poor resilience is further degraded and the effects of the climate catastrophe are amplified.[50]

A significant portion of the world's renewable and non-renewable resources are found in the African continent.[51] Minerals, forests, natural gas, arable land, oil, and animals are all abundant. Over 70% of people living in sub-Saharan Africa depend on forests, woods, and farmland for a living with natural capital making up between 30 and 50% of the continent's overall wealth.[52] While a sizeable amount of these resources is used in an unsustainable manner, some resources are reduced by illegal activity.

The GEO-6 Regional Assessment for Africa reports that the environment is degrading more quickly than anticipated. To estimate illicit wildlife trafficking, forestry, mining, unrestricted fishing, and environmental degradation, cause $195 billion in yearly losses to Africa's natural capital.[53] This illustrates the urgent need for governments to take action to stop the trend. Unquestionably, the news media must help set the agenda for environmental concerns if they are to receive the necessary attention. They can achieve this through educating the public, spreading awareness, generating funds, and bringing attention to environmental problems. Unfortunately, in the majority of sub-Saharan African nations, environmental concerns are either not handled at all or are inadequately covered.[54]

4 Research Methodology

The paper make use of documents and literature reviews from both private and government authorities and is Exploratory in its emphasis. The aim is to bring to our consciousness areas that we need to think about on the other hand, descriptive

[48] Kotir [22].

[49] Thornton [23].

[50] Agyeman [21].

[51] Adedoyin [24].

[52] Inglesi-Lotz [25].

[53] Adedoyin [24].

[54] Agyeman [21].

research was also used because it tries to document observations of specific events while being less interested in their causes and more focused on gathering information to provide detailed descriptions.[55,56]

5 Results

The argument over environmental justice is also heavily influenced by social justice philosophy. Since it is equally concerned with the equitable allocation of (environmental) resources, authority, and decision-making control, environmental justice is fundamentally a subset of social justice.[57] Those who support environmental justice contend that everyone has a right to equitable access to environmental benefits and protection against environmental "bad," regardless of their demographic makeup [27].[58]

6 Cases and Experiences

This section of the study discusses environmental justice in urban and informal settlements in Zimbabwe, environmental justice and mineral protection in sub-Saharan Africa and the protection of tourism-based areas. Environmental and ecological patterns are created in living settings and will without a doubt have a significant impact on the quality of life of individuals who live there, and this will vary depending on social equality.[59,60,61] Low socioeconomic communities in South Africa, that are also predominantly African, are characterized by an absence of effective waste collection and sanitation services, a lack of access to energy sources, water, and enough nourishment, and by high levels of violence.[62]

---

[55] Strydom, Herman. "An evaluation of the purposes of research in social work." *Social Work/ Maatskaplike Werk* 49, no. 2 (2013).

[56] Marlow [26].

[57] Patel [10].

[58] Hobson [27].

[59] Patel [10].

[60] Heynen, Nikolas C. "The scalar production of injustice within the urban forest." *Antipode* 35, no. 5 (2003): 981.

[61] Jackson, Laura E. "The relationship of urban design to human health and condition." *Landscape and urban planning* 64, no. 4 (2003): 191.

[62] Scott [28].

6.1 *Environmental Justice in Urban and Informal Settlements*

In Zimbabwe, Masvingo's informal settlement has been excluded from the city's waste management system [29].[63] They are unable to collect rubbish, for example, and this situation cannot continue. The ecosystem is being polluted by waste that is carelessly thrown on river banks, roadside ditches, and open places. Sustainable housing does not include environmental pollution. In Masvingo and Harare, due to severe pollution, the water bodies can no longer support aquatic life.[64] People are residing in unsanitary conditions that are characterized by poor waste management procedures. Due to the city of Masvingo's failure to include informal settlements in its garbage collection schedule, waste collection services are nonexistent in these areas. People living in informal settlements have been forced by this to use crude methods of garbage management. These crude methods include recycling, dumping, and rubbish burning.

However, these primitive methods of waste management increase environmental issues. For instance, burning garbage worsens the effects of climate change. Therefore, the local efforts used by residents of informal settlements are making the city's environmental crisis worse. Those who carelessly discard their trash harm the environment, while those who pile it up encourage the growth of pests that spread disease including rats and mosquitoes.[65] This is not sustainable and once again increases the danger to the liveability of their spaces.

6.2 *Environmental Justice and Mineral Resource Exploitation*

Many of the country states in sub-Saharan Africa have borders that are primarily a result of European colonization,[66,67,68] Additionally, post-colonial republics frequently maintained the extensive control and authority over populations that the colonial powers exercised. Many times, the state has been imposed and democratic procedures have been tampered with to advance the power of current rulers or certain tribes. Even in cases when the government changes, people who were previously marginalized by the ruling class sometimes adopt similar traits while in charge (in what may be referred to as the "our turn to eat" phenomenon"). These countries that include Zimbabwe, the Democratic Republic of the Congo and Uganda; have the

[63] Chigwenya [29].

[64] Chigwenya [30].

[65] Ibid.

[66] Nyamwanza [3].

[67] Menton [4].

[68] Isgren [5].

worst levels of insecurity because of their poor institutions, rapacious dictatorships, and pervasive corruption.[69]

Sub-Saharan African weak governments are unable to efficiently manage resources or set policies in place to support long-term social progress.[70] The socioeconomic conditions that arise marginalise a large portion of the population and foster an environment that is conducive to crime and violence. Poor governance in Zimbabwe has led to extraordinary rates of inflation reaching as high as 11.2 million percent. Over 4,000 people died between August 2008 and March 2009 as a result of a cholera outbreak brought on by the failure of water and sanitation management in major cities, and as of this writing, over 5 million people are currently experiencing severe food shortages in Zimbabwe.

Neo-patrimonial practices ensure that tiny, powerful cliques with strong stakes in maintaining the status quo and relatively little consideration for the wider population arise where resources are used for wealth accumulation.[71] Unfortunately, foreign funding helps to keep these wealthy elites in power. The myriad NGOs and assistance organizations have grown too fragmented, yet they nevertheless unwittingly support governments by providing the services they ignore. During the Cold War, the aid complex was frequently criticized for being overly centralized and for supporting corrupt regimes. Donor nations in turn offer the centralized support necessary to keep such governments in power.[72]

Negative effects also result from foreign countries and businesses' interest in the riches of Africa. There is now more competition for scarce resources as a result of China's expansion and the rising needs of its middle class for consumer items and automobiles.[73] Peak oil worries have exacerbated what has been referred to as a worldwide "race for resources," with much of this scramble occurring in Africa. Chinese commerce with Africa increased nearly double between 2002 and 2003 to reach $18.5 billion; by 2007, it had risen to $73 billion. China has courted sub-Saharan national governments with a mix of diplomacy and unconditional loans and subsidies to win business for Chinese corporations.[74]

Although China's interests in sub-Saharan Africa have received a lot of attention, other nations' and multinational firms' interests have also played a role in problems with resource management.[75] Too frequently, multinational corporations' operations in Africa focus exclusively on extractive industries in addition to supporting corrupt regimes. The area is viewed as a source of raw materials that are shipped to be processed elsewhere, so erasing the additional value produced by the industrial and service sectors. Due to a lack of investment interest from international companies,

---

[69] Ibid.

[70] Cheru, Fantu. "Structural adjustment, primary resource trade and sustainable development in sub-Saharan Africa." *World Development* 20, no. 4 (1992): 497–512.

[71] Nganje [31].

[72] Irwin [32].

[73] Sheth [33].

[74] Irwin [32].

[75] Sheth [33].

the growth of these secondary and tertiary sectors in the region is hindered. The state ironically is funded by the existing extractive type of investment and is largely to blame for the conditions that lead to this lack of investment. Additionally, natural resource income from being put into more long-term sustainable sectors has been hindered by the short-termism and kleptocratic inclinations of many regimes.

As a result, most of sub-Saharan Africa continues to have some of the poorest populations in the world, despite having a plethora of natural resources.[76] The need to solve these challenges will become increasingly more pressing as populations grow. Without immediate mitigation measures, the effects of climate change, which will have a particularly negative impact on sub-Saharan Africa and have significant implications for the viability of agriculture (both in the commercial and subsistence sectors), could exacerbate current issues to the point of an unprecedented humanitarian catastrophe. Tragically, because of the present misuse of resources and other causes, poverty is so pervasive in the region that many people do not perceive the effects of climate change to be of urgent importance. Sub-Saharan Africa must undergo considerable development and climate-proofing of infrastructure, agriculture, and industry if it is to weather the approaching climate calamity.[77] To balance the region's resource availability with the growing resource demands from the developed world, governments must make sure that the advantages are both sustainable and widely distributed.

The Environmental Protection Agency (EPA) evaluated the effects of natural resource extraction on the environment in the United States. To recent statistics on air pollution in the U.S., the extraction of minerals is responsible for around 30% of the emissions of Pb, 25% of PM, 18% of SO*, 13% of volatile organic compounds (VOC), 3% of CO, and 2% of NO.[78] Mining is the most significant source of PM, smelting is the most significant source of SO*, and processing of crude oil and natural gas is the most significant source of NO*; the three main pollutants to that mineral production contribute significantly. However, the percentage of emissions caused by the mining of mineral resources has dramatically decreased during the past several decades in the United States [35].

Between 1976 and 2000, over 37 000 km^2 or 0.2% of the earth's land surface, was used globally for mining purposes. 40% of disturbed regions are utilized for overburden and other comparable trash disposal, and the remaining 60% are used for excavation.[79] Modern materials including composites, polymers, ceramics, and plastics are made available on the market. These materials frequently compete as alternatives against traditional minerals. By recycling various materials such as glass, metals, some industrial minerals, plastic, etc., environmental issues are also reduced [35]. Recycling is the ideal method of reusing minerals. Because of the numerous ethical issues that the globalization of environmental concerns creates and the need to

[76] Irwin [32].

[77] Ziegler [34].

[78] Sheth [33].

[79] Ziegler [34].

solve them, miners must extract and use minerals in an environmentally sustainable manner.

6.3 Biodiversity Protection in Tourism-Based Areas

On December 29, 1992, the Convention on Biological Diversity went into effect. The development of national policies, programs, or action plans for the preservation and sustainable use of biodiversity is required of the contracting parties.[80] However, the contractual parties have complete control over the structure and substance of any strategies, programs, or plans. The agreement makes a point of stating that the signatory countries have sovereign rights over biodiversity on their territories. Although this is mitigated by mentions of the obligations that governments have to one another given that biodiversity is a shared concern, the contracting parties have the option to decide on the amount of protection to be provided based on how serious they view the issues that biodiversity loss poses.[81]

If the aforementioned suggestions are followed, the Transfrontier Conservation Area (TFCA) in Sub-Saharan Africa can aid in both the preservation of biodiversity and the reduction of poverty.[82] Experience has shown that most governments struggle to implement their policies, particularly those that encourage community involvement in biodiversity protection and tourism growth. This also applies to land reform; outside of a few anecdotal cases (such as in South Africa and Namibia), rural poor communities have not benefited from land reform policies and legislation because many Southern African countries have developed land reform legislation that has been poorly implemented by the governments.

To successfully tap into the expanding tourist business, political stability and the supply of necessary infrastructure in rural areas are essential. Most governments tend to concentrate their efforts on developing metropolitan regions' infrastructure, which limits the potential for tourism growth and reduces chances for rural residents to participate in and make a livelihood from tourism promotion and development.

On the other hand, there is an implicit assumption that the expansion of tourism will continue to be the primary driver of economic growth and that southern African nations would greatly benefit from creating and promoting rural tourist destinations to increase their ability to compete in the international tourism market.[83] Failure to do so would result in no markets for the "products," that are animals, and as a result, the rate of return on conservation would be inadequate to compete with the competitive

---

80 Šeparović [36].

81 Nobs [37].

82 (ibid).

83 Rogerson, Christian M., and Jayne M. Rogerson. "Africa's tourism economy: Uneven progress and challenges." The Routledge handbook of African development (2018): 545–560.

rate of return on alternative land uses, such as unsustainable agriculture and livestock husbandry, hastening the environmental and land degradation in marginal areas.[84]

The mismatch between government policy development procedures and the academic discussion of the relationships between biodiversity protection and poverty reduction is another issue in Southern Africa.[85] To make sure that all biodiversity conservation programs achieve their main goals of preserving ecosystem integrity, conserving biodiversity, and directly assisting in the reduction of poverty, mechanisms to close this gap and mean and capacity for evaluation of all biodiversity conservation programs should be developed.[86]

7 Discussion

The movement for environmental justice has mostly been a US phenomenon. Environmental equality and justice are sometimes used interchangeably with the phrase "environmental racism." Statistics demonstrate that race is a stronger indication than money in assessing the likelihood that a town is contaminated, to some in the US context.[87] Benjamin Chavis, the former chairman of the United Church of Christ's Commission on Racial Justice coined the term "environmental racism".[88] Race was consistently a more important influence in the site of commercial hazardous waste facilities than any other component analysed to research done by the Commission titled Toxic Wastes and Race in the United States quoted in Collins dictionary.[89] When applied to this situation, environmental racism is taken as "an extension of racism in the household, land use, employment, and educational policies and, as such, as a component of the greater web of institutionalized racism" Collinsdictionary.[90]

Ideally, there should be widespread community participation in risk assessment procedures and the post-event diagnosis of purported injuries when it comes to mining and potential environmental impact. Investigation of purported hazards must take into account a wide range of stakeholder interests and viewpoints if the evaluation of toxicity, pollution, and contamination is to be free from corrupt procedures. From an affirmative standpoint, research has shown, that involvement is crucial for the legitimacy of environmental decision-making and for improving problem-solving.[91,92,93]

[84] Šeparović 36.

[85] Munthali [38].

[86] Rogerson, Christian M., and Jayne M. Rogerson. "Africa's tourism economy: Uneven progress and challenges." *The Routledge handbook of African development* (2018): 545–560.

[87] Collin [39].

[88] Cutter [40].

[89] Collin [39].

[90] Ibid.

[91] Scott [41].

[92] Scott [41].

[93] Steele [42].

Individuals should be involved as deliberators and contributors in their own right if sustainability is the aim, prudence necessitates considering numerous options, and community engagement is to be beneficial.[94]

However, when environmental harm is disputed—conceptually and empirically—and there are significant specific social interests at stake (governments, businesses, workers, consumers, environmentalists, residents), those in positions of power often shape public debate in ways that reduce participation and deliberation.[95,96] Furthermore, it is uncommon for scientific proof to be unquestionable and for environmental harm to be proven by merely letting the data speak for itself. The definitions of "science," "evidence," "scientific expert," and "sensible" public policy, for instance, are all influenced by factors like the economy, the scientific tradition within a given national context, the scientific standards that are applied to particular issues, and the style and mode of government.[97] This applies to other environmental conflicts as well, such as those involving mining and community poisoning.

Additionally, there is more room to connect the idea of "social license to operate" with environmental justice concerns more generally (SLO). By applying SLO principles critically, the aim of environmental justice can be partially addressed realistically. Industry officials are already aware of how critical it is to establish some kind of social legitimacy by cooperating with the community and incorporating this into current business practices.[98] Commentators caution that the SLO must be more than a façade for business as usual, a fictitious plan to win over the community, and/or not be overly risk-focused at the price of cooperation.[99,100]

They contend that a strong SLO can be measured, requires effort, and necessitates a desire to interact with stakeholders. Raising the bar for a social license might open doors for forming connections that guarantee more egalitarian and long-lasting consequences. The project of environmental justice depends on this. Without immediate action to address climate change in Africa, the current vulnerability will only worsen, with over 85 million people in the region migrating or being displaced more than in any other region in the world, and an additional 40 million people in sub-Saharan Africa being forced into extreme poverty. Development-proofing Africa's climate initiatives are crucial if there is any chance of tackling the climate issue on the continent before COP27. Africa's unique vulnerabilities must be fully taken into account for this, and climate solutions must be centred on them.

---

94 White [6].

95 Cleary [43].

96 Munro, Iain. "The management of circulations: Biopolitical variations after Foucault." *International Journal of Management Reviews* 14, no. 3 (2012): 345–362.

97 White [6].

98 Lacey [44].

99 Boutilier, Robert G., L. Black, and Ian Thomson. "From metaphor to management tool: how the social license to operate can stabilise the socio-political environment for business." *International mine management 2012 proceedings* 227 (2012): 237.

100 Owen, John R., and Deanna Kemp. "Social licence and mining: A critical perspective." Resources policy 38, no. 1 (2013): 29–35.

The Sendai Framework for Catastrophic Danger Reduction (DRR) 2015–2030 offers nations specific steps they may take to safeguard their development objectives from the risk of disaster occurrences. To avoid the emergence of new risks, lower already-existing risks, and boost resilience, it focuses on methods that address exposure to hazards, vulnerability and capacity, and hazard characteristics. Targets for SDG1 (No Poverty), SDG11 (Sustainable Cities and Communities), and SDG13 all entail adopting and implementing disaster risk reduction policies following the Sendai Framework (Climate Action). Only information on 30 African nations' adoption and execution of Sendai-compliant disaster risk management plans is available. Data are unavailable for four of the ten nations most susceptible to climate change: Chad, Guinea-Bissau, DR Congo, and Mali. African nations generally do a little worse than the global average (0.61 versus 0.64, respectively). On the continent, Cameroon and the Comoros have the lowest scores, while Algeria and Ethiopia have the highest. This makes it critical to see how environmental justice can be applied in real world contexts, to improve equity among nations.

8 Conclusion and Future Direction

Habitat-based conservation techniques work better than species-based ones. Species may be conserved even if they are not appreciated for their reason because of how they help other species survive. Likewise, as ecological services are by-products of habitat, they might be safeguarded. The preservation of habitat benefits watersheds, carbon storage, the collection of non-timber forest products, recreational usage, and scientific research. However, even while preserving wild plants and animals in nature or wildlife reserves, national parks, zoos, arboreta, or botanical gardens is an essential short-term measure when such resources are in danger of going extinct immediately, it does not solve the larger issue of biodiversity loss. The preservation of all natural resources in a small number of wildlife reserves should not be the aim of biodiversity conservation. Critical biodiversity thresholds should be safeguarded globally.

The establishment of secure land tenure/use rights, the removal of subsidies that encourage land conversion, the development of mechanisms to transfer the benefits of wildlife use to local communities, and the establishment of access to credit/real assets, technology, and markets to open up alternative economic opportunities are examples of the last three. In other words, a strategy for biodiversity conservation should attempt to safeguard crucial ecological thresholds.

It should align the social and private costs of using biological resources and provide resource users with the tools they need to address the social costs in a way that serves societal objectives. It is insufficient to charge resource users what environmental resources are worth. Additionally, they must be able to react correctly. In the end, whatever happens to this planet is all our fault as individuals and as a species. But because individuals and groups of individuals have contributed in diverse ways over time, both to environmental good and ill, the accompanying obligations, particularly for damages, must likewise not be the same and equal but rather differentiated, justly differentiated, if justice is to be served.

To guarantee that resilience and equality objectives are met, climate change policies and initiatives need to be examined and modified over time. To guarantee that funds reach target populations and provide equitable resilience results, regular monitoring and assessment using fair indicators and progress metrics, budgetary openness, and equitable resource allocation plans are crucial. The country as a whole is included in the definition of "interested and affected parties" for mining in protected areas, thus mining corporations must make sure that they conduct extended meaningful engagement with all stakeholders. The researcher makes the argument that it should be illegal to mine in protected areas since they are essential to maintaining Africa's status as a top tourist destination and protecting biodiversity, animals and vital natural ecosystems. Furthermore, governments should develop strategies to distribute the benefits to the local community while supporting the maintenance of protected areas. Ecotourism should be encouraged in protected regions known for, among other things, their exceptional beauty and ecological relevance.

Environmental justice has emerged as a key component of this reform agenda. New development policies, laws, and institutions are currently being developed by governments and civil society movements all over the world to lessen or prevent the disproportionate effects of growth on the poor while accelerating the advantages that can result from sustainable resource use. Communities and governments are using environmental justice as a strategy to address the social dimensions of the problem that have contributed to divergent development outcomes between the rich and poor, between and within countries, and the distributional effects of toxicity on the weak and vulnerable.

Understanding the contextual, political-economic challenges that governments and communities must overcome to respond to the need for more sustainable and equitable development is crucial. More and more nations in the Sub-Saharan are realising the importance of improved legal structures and procedures for managing trade-offs and fostering triple-wins in terms of economic, social, and environmental concerns. Another important lesson is that adapting governance solutions to various social, normative, and ecological settings is essential to the efficacy of such policies. Awareness of the future of environmental justice principles will need an understanding of the growing synthesis between global and local norms and systems.

References

1. Schaefer A (2022) Is justice a fixed point? Am J Polit Sci
2. Botterhuis L, van der Duin P, de Ruijter P, van Wijck P (2010) Monitoring the future. building an early warning system for the Dutch Ministry of Justice. Futures 42(5), 454–465
3. Nyamwanza AM, Kujinga KK (2017) Climate change, sustainable water management and institutional adaptation in rural sub-Saharan Africa. Environ Dev Sustain 19:693–706

4. Menton M, Larrea C, Latorre S, Martinez-Alier J, Peck M, Temper L, Walter M (2020) Environmental justice and the SDGs: From synergies to gaps and contradictions. Sustain Sci 15:1621–1636
5. Isgren E, Andersson E (2021) An environmental justice perspective on smallholder pesticide use in Sub-Saharan Africa. J Environ Dev 30(1):68–97
6. White R (2012) Environmental forensic studies and toxic towns. Curr Issues Crim Just 24(1):105–119
7. Bullard RD (ed) (2005) The quest for environmental justice: Human rights and the politics of pollution, vol 19. Sierra Club Books, San Francisco
8. Bullard RD (1994) Overcoming racism in environmental decisionmaking. Environ Sci Policy Sustain Dev 36(4):10–44
9. Pellow D (2007) Naguib. Transnational movements for environmental justice. Mit Press, Resisting global toxics
10. Patel Z (2009) Environmental justice in South Africa: Tools and trade-offs. Soc Dyn 35(1):94–110
11. Rogge ME (2008) Environmental justice. In: Encyclopedia of social work. http://socialwork.oxfordre.com.ez.sun.ac.za/view
12. Douglas I, Alam K, Maghenda M, Mcdonnell Y, McLean L, Campbell J (2008) Unjust waters: climate change, flooding and the urban poor in Africa. Environ Urban 20(1):187–205
13. Agyeman J (2014) Environmental justice and sustainability. In: Handbook of sustainable development. Edward Elgar Publishing, pp 188–205
14. Chunn D, Boyd S, Menzies R (2002) We all live in Bhopal': criminology discovers environmental crime. In: Toxic criminology: environment, law and the state in Canada. Fernwood Publishing, Halifax, pp 7–24
15. Saha R, Mohai P (2005) Historical context and hazardous waste facility siting: understanding temporal patterns in Michigan. Soc Probl 52(4):618–648
16. Williams C (1996) An environmental victimology. Soc Justice 23(4):16–40
17. Rush S (2002) Aboriginal resistance to the abuse of their national resources: the struggles for trees and water. In: Boyd S, Chunn D, Menzies R (eds) Toxic criminology: environment, law and the State in Canada. Fernwood Publishing, Halifax, pp 67–86
18. Engel S, Martin B (2006) Union Carbide and James Hardie: lessons in politics and power. Glob Soc 20(4):475–490
19. Brook D (1998) Environmental genocide: native Americans and toxic waste. Am J Econ Sociol 57(1):105–113
20. Millner F, Ruddock K (2011) Climate litigation: lessons learned and future opportunities. Altern Law J 36(1):27–32
21. Agyeman J, Bullard RD, Evans B (2012) Introduction: joined-up thinking: bringing together sustainability, environmental justice and equity. In: Just sustainabilities. Routledge, pp 19–34
22. Kotir JH (2011) Climate change and variability in Sub-Saharan Africa: a review of current and future trends and impacts on agriculture and food security. Environ Dev Sustain 13:587–605
23. Thornton PK, Jones PG, Owiyo T, Kruska RL, Herrero MT, Kristjanson PM, Notenbaert AMO, Bekele N, Omolo A (2006) Mapping climate vulnerability and poverty in Africa
24. Adedoyin FF, Ozturk I, Agboola MO, Agboola PO, Bekun FV (2021) The implications of renewable and non-renewable energy generating in Sub-Saharan Africa: The role of economic policy uncertainties. Energy Policy 150:112115
25. Inglesi-Lotz R, Dogan E (2018) The role of renewable versus non-renewable energy to the level of CO_2 emissions a panel analysis of sub-Saharan Africa's Big 10 electricity generators. Renew Energy 123:36–43
26. Marlow CR (2023) Research methods for generalist social work. Waveland Press
27. Hobson K (2004) Environmental justice: an anthropocentric social justice critique of how, where and why environmental goods and bads are distributed. Environ Polit 13(2):474–481
28. Scott DN (2005) Shifting the burden of proof: the precautionary principle and its potential for the "democratization"of risk. Law Comm Can ed 50–85

29. Chigwenya A, Simbanegavi P (2021) Towards sustainable housing: waste management in informal settlements in Masvingo City, Zimbabwe
30. Chigwenya A, Wadzanai P (2020) The intersectionality between right to the city, informality and waste management in Masvingo City, Zimbabwe. J Urban Syst Innov Resil Zimbabwe-JUSIRZ 2(2):26–52
31. Nganje F (2015) Moving beyond Africa's crisis of institutions
32. Irwin DA (2020) Free trade under fire. Princeton University Press
33. Sheth JN (2011) Impact of emerging markets on marketing: Rethinking existing perspectives and practices. J Mark 75(4):166–182
34. Ziegler CE (2006) The energy factor in China's foreign policy. J Chin Polit Sci 11:1–23
35. EPA, US (1991) Guideline for regulatory applications of the urban airshed model. US Environmental Protection Agency, Research Triangle Park, NC (EPA-450/4-91-013)
36. Šeparović M, Mrčelić GJ, Slišković M (2014) The Implementation of the convention on biological diversity. In: The 6th international marine science conference, p 201
37. Nobs B, Lamb R, Pythoud F(2003) The cartagena protocol on biosafety of the convention on biological diversity: overview and future development. Methods Risk Assess Transgenic Plants: IV. Biodivers Biotechnol 93–99
38. Munthali SM (2007) Transfrontier conservation areas: integrating biodiversity and poverty alleviation in Southern Africa. In: Natural resources forum, vol 31, no 1. Blackwell Publishing Ltd., Oxford, UK, pp 51–60
39. Collin RW (1993) Law: environmental equity and the need for government intervention: two proposals. Environ Sci Policy Sustain Dev 35(9):41–43
40. Cutter SL (1995) Race, class and environmental justice. Prog Hum Geogr 19(1):111–122
41. Scott DN (2005) When precaution points two ways: confronting "west nile fever." Can J Law Soc/La Revue Can Droit et Soc 20(2):27–65
42. Steele J (2001) Participation and deliberation in environmental law: exploring a problem-solving approach. Oxf J Leg Stud 21(3):415–442
43. Cleary P (2012) Mine-field: the dark side of Australia's resources rush. Black Inc.
44. Lacey J, Parsons R, Moffat K (2012) Exploring the concept of a social licence to operate in the Australian minerals industry

Control of Production, Consumption and Disposal of Plastic: Urban Sustainability and Human Health

Shahnaz Tabassum, Raman Jasrotia, and Seema Langer

Abstract Plastics have become an important part of our modern society due to its flexibility, diverse function and also because it is budget friendly. But it is also one of the major threats to the sustainable urban environment and human health as it is used extensively in the present time. This chapter encompasses some of the core issues of plastic production, consumption and plastic waste, and discusses ways to control it to minimize its negative impact on the environment. Two main harmful consequences of plastic production are depletion of natural resources and also the release of hazardous chemicals while it is prepared in the industries. The consumption of plastic especially the single use plastic leads to generation of plastic waste in a huge amount thus keeping a check on the plastic consumption and finding a safe and less toxic alternatives is another major issue to be considered. And finally, the proper management of disposal of the plastic waste is required. If not disposed in a proper manner these plastics persists for centuries and pollute the air, soil and water leading to the disruption of the food chain and resulting in degraded human health linked to disruption of the endocrine system and cancer etc. Thus, a proper regulation on the production, consumption and disposal of plastic is required via, increased public awareness, research and making collaborative policies by international organization.

Keywords Pollution · Plastics · Endocrine disruptors · Disposal · Urban management · Sustainability

S. Tabassum · R. Jasrotia (✉) · S. Langer
Department of Zoology, University of Jammu, Jammu 180006, India
e-mail: drraman.jasrotia@jammuuniversity.ac.in

R. Jasrotia
Department of Zoology, Cluster University of Jammu, Jammu, India

J. A. Parray et al. (eds.), *Sustainable Urban Environment and Waste Management*, Urban Sustainability, https://doi.org/10.1007/978-981-96-1140-9_9

1 Introduction

The Stone, Bronze and Iron age is followed by the modern times of the plastic age. With the creation of celluloid, in the latter part of the nineteenth century, Plastic is originated, using modified cellulose. It was formed aiming to replace natural materials like horns, ivory etc. With time, Bakelite was invented which was the first thermostat plastic. By 1930s various other synthetic plastics were synthesized and being used as the understanding of polymer had advanced. In 1960s, the knowledge of new polymers, like fluoropolymers, polycarbonate expanded the range of its application. Plastic replaced leather, metals in various areas of our lives and even the automobile and the aerospace industries.

Now in the modern society, plastic has become an indispensable part, transforming our daily lives and the industries. Its versatile utility, the durability and low production costs have made it an integral part of our everyday life. It is almost used and found in every corner of the modern world like from packaging a very tiny item, construction of large buildings and in large chemical industries. However, its extensive use has led to significant challenges, surrounding the environment and human health. As cities keeps on emerging and with the increase in population it is a must to keep a proper check on the plastic production and its management, so that it doesn't affect the sustainable development of these region.

The term Urban sustainability refers to the capacity or ability of a city to function in a way that leads to a long-term well-being of the people residing in those areas and giving them the maximum facilities with minimum environmental impact. It includes a wide range of factors including the judicial use of resources waste management environmental protection and the well-being of the population.

As mentioned earlier, the population shifts from rural to urban area is taking place, with more than half of the world's population presently living in the cities, thus focusing on the urban sustainability has become the need of hour and one of the important pillars to achieve this is protecting the environment which encompasses minimizing the pollution and maintaining biodiversity. The tremendous effect that plastic has on lives and the environment highlight how urgent it is to address these issues. Like if we look into the health implication by the use of plastic extends beyond its effect on the environment. The harmful substances released by the disposal of plastics include bisphenol A (BPA), phthalates which are potent endocrine disruptors. These substances have the ability to seep into food and drinks items packaged in plastic, which could be harmful to people health. Numerous health problems, including hormone abnormalities, reproductive trouble, developmental disorder have been related in studies to exposure to these chemicals. Furthermore, the health of humans is becoming increasingly concerned about microplastic particles smaller than five mm. These particles are result of the bigger plastics breaking down. In order to tackle these issues related to plastic, a comprehensive comprehension of its environmental and health effect is necessary, in addition to well thought plan to manage its manufacture, usage and disposal. To lower its harmful effect, it is very important to focus on controlling the manufacturing of plastic. This includes

Fig. 1 Plastic waste generation and future projection

imposing restrictions on the kind and amount of plastic that can be produced as well as promoting the creation of substitute material that pose fewer environmental risks.

Controlling the Consumption of plastic is equally vital. Reducing plastic wastes can be achieved in significant ways by informing customers about the negative environmental effect of plastic and encouraging the use of reusable products. Lastly handling the plastic that does become waste requires efficient disposal technique. To guarantee that plastic trash is treated effectively and sustainably, recycling programs and waste segregation technique must be improved (Fig. 1). We can greatly lessen the environmental impact of the plastic by controlling its manufacture, use and disposal through targeted actions.

I. **Control on Production of Plastics**:

One of the purposes for which plastic is frequently used is packaging. Out of the total plastic manufactured 40% of plastic material are utilized to stock material and packaging things in different companies [7]. Plastic's adaptability has made it possible to package food items like bread, rice, milk and a variety of pharmaceuticals in an efficient and economical manner. These items have a significant environmental post utilization waste load because of their growing number of its application. The increase demand has also led to the increase production which has led to various problems.

2 The Problem with Plastic Production

(1) **Human Health Implications of plastic production**:

Chemical hazards:

The impact of plastic production on health is caused by the chemicals used during the production of plastic. Some of the main chemicals are bisphenol A, Phthalates and some other endocrine disruptors, leads to dangerous health risk in a long run. Phthalates can escape from products because they are not chemically integrated into the plastic matrix resulting in their release into the environment. The widespread use and large-scale production of these chemicals have drawn significant scrutiny [22]. Phthalates and BPA have been detected in aquatic environment, airborne dust and even in the atmosphere due to their volatile properties [16, 17]. The potential harmful effect of these substance on both wildlife and human health have waste considerable alarm.

Furthermore, some researchers have indicated that BPA may elevate the risk of breast cancer, prostate cancer, pain and metabolic disorder. In women exposure to BPA has been linked to various health issues including obesity, endometrial hyperplasia, recurrent miscarriages, infertility and polycystic ovarian syndrome [14, 23]. The concentration of unconjugated BPA in human blood and tissue typically ranges from 0.1 to 10 micro-liter and the level are mentioned in blood serum and urine [19].These hazardous chemicals can sweep down into the food or beverage causing all these health issues.

Microplastic

The microplastics can also enter the human body indirectly through personal care products like facial cleanser, soaps, toothpaste and exfoliating scrubs. Studies have indicated that a considerable percentage of these items contain micro beads—50% in face wash and 67% in facial scrubs [3]. The presence of these microbeads can exacerbate Skin ageing and leads to the development of dark spots by creating microscopic gaps that allow bacteria to penetrate in the skin. These microplastic also have hazardous impact on the human health leading to gastrointestinal issues resulting resulted due to inflammation of the digestive tract. Due to disruption in the microbiota, a person may suffer from IBS, constipation etc. Recent studies have also proved that microplastic alters the gut bacteria, and affect the immune cells of a human [9]. Few other problems caused by microplastic are respiratory problems. These minute particles can lead to irritation and inflammation in the respiratory tract, resulting in symptoms like difficulty breathing, coughing and aggravation of existing respiratory issues such as asthma [1]. Allergic reaction, skin irritation are a few of other health concerns. Microplastic is even found in the human placenta [13].

Airborne Plastics

Microplastics in the air pose a serious threat to the environment and have an impact on ecological systems and human health. These particles come from a variety of sources, such as industrial pollutants, synthetic textiles, waste management procedures, and agricultural operations. Because synthetic garment fibres shed, indoor microplastic concentrations are typically higher than outdoor ones, which are influenced by weather conditions like wind and rainfall. These microplastics are inhaled, eaten, and come into touch with the skin by humans. The methods for identifying and detecting airborne microplastics are described in this section, with an emphasis on sample preparation, various analytical techniques, and active and passive sampling methods. These methods include scanning electron microscopy with energy-dispersive X-ray spectroscopy (SEM-EDS), Raman spectroscopy, Fourier-transform infrared (FTIR) spectroscopy, and microscopy. They are used to examine the chemical and physical details of the microplastic.

2. **Impact on Environment**:

Land: Toxic chemicals released by chlorinated plastics may seep into the soil and contaminate local water sources, including groundwater, endangering wildlife that depends on them. Microorganisms in landfills, which are continuously stocked with a variety of polymers, hasten the degradation of plastic. Methane, a major greenhouse gas, is produced during this degradation process. Many landfills lack the mechanisms necessary to extract and use methane for electricity, while some do. Even though they break down more slowly, biodegradable polymers can produce methane when left exposed.

Water: During shipping, tiny plastic pellets called "nurdles" are frequently spilt into the water. Nurdles are thought to make up around 10% of beach debris globally. While they don't always decompose completely, plastics in marine environments usually release dangerous compounds like bisphenol A and polystyrene into the ocean over the course of a year. Together with food containers and plastic bags, which make up the majority of maritime waste, polystyrene pieces and nurdles are among the most prevalent forms of plastic debris found in the ocean. It was projected in 2012 that the world's oceans had 165 million tons of plastic waste. This pollution may damage marine life, which could have an impact on the availability of food for humans [15].

Effective efforts to limit plastic manufacturing are essential given the significant hazardous effects of plastics described in the previously, which range from health concerns and environmental pollution to ecosystem deterioration and the aggravation of climate change. The detrimental effects of plastics on the environment and human health highlight the urgent need for all-encompassing solutions to address these problems. Therefore, enforcing strict rules, encouraging sustainable alternatives, improving recycling systems, and funding creative research should be the main priorities of the following actions. In order to address the underlying causes of plastic pollution and guarantee a safer, more sustainable future, the first step is to control its production

International Regulations and legislation.

1. **Basal Convention**: The purpose of the Basel Convention is to regulate the transboundary movement of hazardous wastes, especially plastic waste, and to guarantee theirpropermanagementanddisposal.

   Important Provisions: Prevents illicit dumping and environmental damage by controlling the import and export of plastic trash and requiring appropriate documentation and management.

2. **The Single-Use Plastics Directive of the European Union (EU)** aims to lessen the environmental impact of single-use plastic products.

   Important clauses: It include a ban on single-use plastics like straws, plates, and cutlery, standards for waste management from producers, and goals for lowering plastic pollution in marine habitats.
3. **The goal of the United Nations Environment Programme (UNEP)** Clean Seas Campaign is to combat marine plastic pollution worldwide by promoting awareness, forming alliances, and offering policy recommendations. The aim is to tackle the issue of marine plastic pollution worldwide by means of policy suggestions,

   Important Points: Promotes national measures that cut down on plastic consumption, enhance garbage disposal, and foster global collaboration.
4. **Plastic Waste Partnership (PWP)** Goal: To enhance plastic waste management worldwide, this program is a part of the Basel Convention.

   Important Features: Promotes collaboration amongst corporations, NGOs, and governments in order to create solutions for the management and minimization of plastic waste.
5. **International Maritime Organization (IMO) MARPOL Annex V** Goal: Controls the discharge of plastics and other garbage to prevent marine pollution from ships.

   Important provisions: forbids ship from discharging plastics into ocean and mandates that they appropriately handle plastic garbage on board.
6. **The United Nations Sustainable Developmental Goals** Goals: Uses a variety of objectives to address global issues including plastic pollution.

   Important points: SDGs12 (Responsible consumption and production) and 14 (life below water encourage initiatives to lower plastic waste, enhanced recycling and support sustainable practices).

Commitments by Industries:

One of the most important factors in lowering plastic production is the private sector. Numerous businesses have made a commitment to adopting sustainable practices, like reducing packaging, employing recycled materials, and aiding in waste management programs. Businesses come together to create solutions for plastic pollution through industry alliances like the Alliance to End Plastic Waste.

Alternatives to Plastic:

One of the most important tactics for lowering plastic production thus pollution and boosting sustainable practices is to promote substitutes for traditional plastics. Here are a few detailed illustrations of how this can be accomplished:

1. **Materials that Degrade Biodegradable**

Biodegradable Polymers: Bioplastics, a well-known example of biodegradable materials, are made from renewable resources like algae, sugarcane, and corn starch. For example:

Polylactic acid, or PLA, is a material used in 3D printing, disposable cutlery, and food packaging. It is produced from fermented plant starch. Because PLA can break down in industrial composting settings, it has less of an impact on the environment than traditional plastics.

PHAs, or polyhydroxyalkanoates, are made by bacteria that break down organic materials. They are utilized in packaging and medical equipment, among other things. In both marine habitats and industrial composting, PHAs decompose into innocuous chemicals. Examples:

Danone's Use of PLA: In an effort to lessen its dependency on traditional plastics, Danone, a well-known food and beverage firm, has started utilizing PLA in part of their yogurt packaging.

2. **Products that can be Reused**

Reusable Substitutes: Reusable product substitution is another practical strategy for cutting down on plastic waste. As examples, Reusable Bags: Businesses and governments promote the use of reusable shopping bags composed of recycled plastic, jute, or cotton. For instance, the use of reusable bags has surged as a result of the adoption of plastic bag bans in locations likeNewYork.

Reusable water bottles: By using glass or stainless-steel bottles, you can cut down on the number of single-use plastic bottles you need. These alternatives are promoted by companies like S'well and Hydro Flask, who frequently provide consumers with rewards like loyalty plans or discounts.

Zero-Waste Shops: Offering a variety of reusable and eco-friendly products, shops such as Package Free Shop and Zero Waste Home assist customers in cutting back on their use.

3. **Monetary Rewards**

Tax breaks and subsidies: The transition to sustainable alternatives can be accelerated by offering financial incentives to businesses who manufacture or utilize eco-friendly products. As examples, consider:

Government Grants: Governments have the authority to provide grants or subsidies to businesses who invest in environmentally friendly packaging solutions or are creating novelbiodegradable materials.

Tax Benefits: Businesses that use environmentally friendly techniques may be eligible for tax exemptions or breaks. For instance, in several nations, companies that make investments in sustainable Lower tax rates can be advantageous for materials.

Real-World Examples:

European Union Green Taxonomy: The EU offers cash rewards to companies that participate in environmentally friendly initiatives, such as those aimed at cutting down on plastic waste.

California's Plastic Pollution Reduction Program provides financing for research and development of alternative materials, among other forms of financial support, to businesses who are coming up with creative ways to cut down on plastic pollution.

Through the promotion of biodegradable materials, reusable items, and financial incentives, communities can effectively decrease their dependence on traditional plastics, thereby making a positive impact towards a sustainable future.

Encouraging Innovation and Research:

The development of novel materials, technologies, and waste management strategies depends heavily on investments in research and innovation. Research into recycling methods, biodegradable polymers, and substitutes for plastics derived from fossil fuels should be funded by both public and commercial sources.

Maintaining human health and attaining urban sustainability depend heavily on managing the production of plastics. Due to the widespread nature of plastic pollution, a multimodal strategy including international collaboration, community participation, technology advancements, and regulatory measures is required. We can build resilient, healthy urban ecosystems and protect the health of present and future generations by emphasizing sustainable activities and minimizing dependency on plastics.

II. **Control on Consumption of Plastic**

Production efficiency is vital, but cutting back on the usage of useless product can be difficult, frequently because of convenience and food safety concern [4]. But the second important step in managing plastic pollution after the control on production is to check the consumption of produced plastics. It is still possible nevertheless, to reduce needless packing by for example, avoiding double packing and choosing more ecofriendly alternatives. One long term strategy to effectively reduce plastic consumption is to increase public understanding of the environmental effect of consumer choices. Both official education in school and unofficial methods, such as news articles and neighbourhood clean-up days, can accomplish this. For instance, it can make a big impact to inform consumers about the advantage of selecting microbeads free alternatives, which might also be reinforced by precise and reliable labelling [6]. Businesses will have to revamp their product lines as consumer demand for plastic free alternative grows [2]. Nevertheless, creating workable substitute is still difficult in necessitates approaches that promote and foster creativity in this field.

The amount of plastic waste produced every year is 400 million tonnes. By 2050, it is predicted that the world's primary plastic production will have reached 1,100 million tons, provided that historical growth rates persist. Concerningly, there has been a trend toward single-use plastic goods—items designed to be discarded after just one brief usage.

A little over 36% of all plastics manufactured are used for packaging, including single-use items like drink and food containers, of which 85% end up as uncontrolled garbage or in landfills. In addition, fossil fuel, or "virgin" feedstock, is used to make about 98% of single-use plastic products. It is anticipated that the amount of greenhouse gas emissions linked to the creation, consumption, and disposal of traditional plastics derived from fossil fuels will increase to 19% of the world carbon budget by 2040 (UN, environment program)

3 Types of plastics [8]

To effectively reduce plastic consumption, it is essential to understand the types of plastics we encounter in our daily lives and the potential harm each type of plastic possess. If we are more informed about the various plastic categories, we can make more wise decisions while using each type of plastics. In response to growing concerns from recyclers across the country, The Society of the Plastics Industry, Inc. (SPI) introduced a resin identification coding system in 1988. This system categorizes plastics into seven distinct classes, each marked with a specific number ranging from 1 to 7, typically found within a recycling symbol on plastic products. These codes help consumers and recyclers identify the type of plastic, its recyclability, and potential health risks. This awareness can guide consumers toward safer and more environmentally friendly options, supporting the overall reduction in plastic consumption and promoting more effective recycling practices.

4 Plastic Made of Polyethylene Terephthalate (PET)

The main application for polyethylene terephthalate, or PET plastic, is the production of single-use water bottles. Additionally, it is used in the production of different utensils and containers for goods including butter, salad dressings, vegetable oils, mouthwash, juices, soft drinks, and cosmetics. PET plastic is distinguished by its smooth, clear, and thin texture. PET is the material of choice for packaging food and beverages because of its anti-inflammatory and liquid-resistant qualities, which provide an airtight seal that keeps oxygen from contaminating the contents. Furthermore, PET bottles are designed to reduce the amount of liquids or beverages that wash out of the inside. Type 1 plastic bottles could contain antimony trioxide even though they don't contain any dangerous microorganisms or phthalates. When antimony is present, it may be carcinogenic. It builds up within the human body. Antimony can leak out of PET containers when they come into constant contact with drinking water, and high temperatures can make this leakage worse. It is essential to keep these bottles away from heat sources as a result. It is crucial to remember that PET plastic is meant to be used just once. PET bottles are therefore usually safe for single use.

4.1 High-Density Polyethylene (HDPE) is Type 2

One of the most extensively used plastics in the world, high-density polyethylene, or HDPE, is made of petroleum and is renowned for its ability to withstand heat. This kind of plastic is widely used to make toys, refrigerators, milk jugs, detergent bottles, and different kinds of plastic grocery bags. HDPE is renowned for its robustness, longevity, and heat resistance. Because it doesn't include any dangerous ingredients like BPA or phthalates, it's a safer option for a range of uses. Nonetheless, studies have demonstrated that long-term contact to sunshine might result in the release of nonylphenol from HDPE because to UV radiation. Type 2 containers are frequently seen to be safer for storing food and beverages than PET containers.

4.2 Polyvinyl Chloride (PVC) is Type 3

Fruit juice and cooking oil are two frequent goods to store in type 3 plastic containers. One kind of plastic in this group that is flexible and simple to Mold is PVC. Nonetheless, it is critical to understand that phthalates, which are hazardous to human health, are frequently used to make PVC flexible. Furthermore, PVC has been linked to a number of harmful compounds, such as cadmium, lead, dioxin, and BPA. There is a decrease in the usage of PVC due to the serious health and environmental dangers associated with its whole lifecycle, from production to disposal. PVC is nevertheless widely used because of its affordability and adaptability. However, the toxicity of PVC presents a number of health dangers, such as cancer, genetic damage, birth defects, chronic bronchitis, ulcers, skin problems, hearing loss, and eyesight impairment.

4.3 Low-Density Polyethylene (LDPE) is Type 4

The heat resistance of low-density polyethylene (LDPE), a polymer generated from petroleum, is well-known. Juice and milk cartons, as well as the packaging of frozen items, are frequently made from LDPE, which is available in both transparent and opaque varieties. Since there is little chance of dangerous materials seeping into the containers or the packaged liquids, LDPE is regarded as safe for the storage of food and beverages.

4.4 *Polypropylene (PP) is Type 5*

PP plastic, or polypropylene, is a strong, semi-transparent polymer that is renowned for its water-repellent qualities, strength, and tolerance to high temperatures. PP is frequently used in the packaging of goods including yogurt, medications, drinks, ketchup, and related items. Research shows that food or water kept in polypropylene containers doesn't contain any dangerous substances. Because they don't offer any health dangers, the majority of PP plastic products are also regarded as safe to use in dishwashers and microwaves. Polypropylene containers are regarded as safe for storing food and beverages, much like LDPE containers.

4.5 *Polystyrene (PS) isType six*

Petroleum-based plastic called polystyrene is made with the well-known carcinogen benzene. Despite being commonly used in insulation and packaging, this material is hazardous to health since it contains styrene. Styrene exposure over an extended period of time has been linked to haematological, cytogenetic, neurotoxic, and cancerous effects. Styrene has been identified by the International Agency for Research on Cancer (IARC) as a possible human carcinogen.

4.6 *Type 7: Additional Polymers*

Polycarbonate and other various plastics not included in kinds 1 through 6 are included in type 7 plastics. BPA, which is frequently present in these materials, can seep into food and drink that are kept in these containers. Since BPA exposure has been associated with health hazards in numerous studies, the use of polycarbonate plastics has significantly decreased. Despite the fact that these plastics are still utilized in consumer goods, type 7 plastics may pose health risks, hence it is becoming less acceptable to use them in items like large reusable water containers and infant bottles.

Use of all these plastics pose severe danger to the human health, environment causing almost every type of pollution, thus various measures should be taken while consuming them or we should try to reduced their consumption as much as possible. The impact of plastic consumption on urban sustainability is profound and multifaceted. One of the most immediate consequences is the increase in waste generation, which has led to overflowing landfills in many urban areas. According to a 2021 report by the World Bank, global waste generation is expected to increase by 70% by 2050, with plastic waste comprising a significant portion. This not only strains waste management systems but also exacerbates pollution in urban environments. Plastics that escape waste management systems pollute streets, rivers, and oceans, where they can harm local ecosystems. For instance, marine life often ingests plastic debris, leading to reduced biodiversity and disrupting food chains. Furthermore, the production and disposal of plastics contribute to climate change, as fossil fuels are required for plastic manufacturing and incineration releases greenhouse gases. The cycle of plastic consumption, waste, and environmental degradation poses a serious threat to the sustainability of urban areas, necessitating urgent action.

The effects of plastic consumption extend beyond environmental degradation, posing significant health risks to urban populations. One of the most concerning issues is microplastic exposure, which has been detected in air, water, and even food sources. A study published in the journal "Environmental Science and Technology" found microplastics in 94% of tap water samples collected globally, raising concerns about human ingestion. These tiny plastic particles can accumulate in the human body, leading to potential health risks, including inflammation and immune system disruption. Additionally, many plastic products contain harmful chemicals, such as bisphenol A (BPA) and phthalates, which can leach into food and beverages. Research has linked these chemicals to various health issues, including hormonal imbalances and reproductive problems. The relationship between plastic waste and disease proliferation is also noteworthy; for example, improperly disposed plastic waste can create breeding grounds for disease-carrying pests, such as mosquitoes, thereby increasing the risk of vector-borne diseases in urban areas. Thus, the health implications of plastic consumption are significant and warrant immediate attention.

5 Regulatory Measures to Control Plastic Consumption

Rules and Regulations

The governing bodies and the citizens together can play in important role in controlling the consumption of plastic. The foundation effective step is the imposing of rules and regulations in the use of plastic. If the government policies are effective than a great impactful reduction in the usage of plastics can be seen. Some of the countries have implemented rules and regulations to reduce the plastic consumption each country has set the rules depending on their specific need and context. Some of the countries with their rules are as follow:

European Union:

The EU has taken the lead in putting comprehensive measures for reducing plastic use into practice:

The Single-Use Plastics Directive (2019/904/EU) prohibits the use of single-use plastic goods like straws, plates, stirrers, and silverware in an effort to lessen the environmental impact of specific plastic products. Additionally, it stipulates that by 2025 and 2030, plastic beverage bottles must have at least 25 and 30% recycled material, respectively.

2018 EU Plastics Strategy: This approach promotes the design and manufacture of plastic items that are more recyclable and sustainable in order to facilitate the shift to a circular economy. It consists of steps to encourage recycling and reuse as well as cut down on plastic waste.

5.1 USA

Each state has its own plastic regulations. Here are a few noteworthy examples:

Regulations for Reducing Plastic Waste in California: The state has put in place a number of regulations, such as the California Plastic Pollution Prevention and Packaging Producer Responsibility Act (SB 54), which mandates that manufacturers decrease the amount of plastic packaging they use and improve its recyclability. Additionally, the state forbids single-use plastic straws at full-service dining establishments.

The city of New York has implemented a ban on single-use plastic bags in retail establishments, with the aim of promoting the usage of reusable bags. Businesses must charge for alternatives like paper bags and punish those who break the law.

5.2 India

India has made a number of efforts to cut back on plastic waste and consumption:

Plastic Waste Management Rules (2016): Under these regulations, the lifetime of plastic items, including collection, segregation, and recycling, must be managed by manufacturers, importers, and brand owners. Additionally, they forbid the use of some single-use plastic products and encourage the usage of recycled materials.

State-Level Plastic restrictions: To promote the adoption of environmentally friendly substitutes, a number of Indian states, notably Maharashtra and Tamil Nadu, have imposed their own restrictions on single-use plastics and products made of polystyrene.

Case Study: Tamil Nadu's Nilgiris Operation Blue Mountain.

Supriya Sahu, the district collector, spearheaded the Operation Blue Mountain campaign in 2001 to outlaw the use of plastic in the district. The famed hill resort of Nilgiris needed the campaign to clear the river sources and springs. The project has been documented as the finest governance practice from Indian States by the UNDP and the former Planning Commission. Images of animals choking were featured in the campaign to help people comprehend. They also discussed how plastic gets into lakes and other bodies of water and clogs drains.

5.3 China

China has implemented important measures to combat the pollution caused by plastics:

China has imposed a progressive prohibition on single-use plastic bags, straws, and other disposable plastic products as of 2020. The rules also seek to encourage the use of biodegradable substitutes and lessen the amount of plastic trash generated by the catering sector.

China forbade the import of plastic garbage in 2018. Prior to this, China was one of the main markets for plastic recycling. Other nations are now compelled to look for alternate ways to handle their plastic garbage.

5.4 Kenya

Kenya has addressed the issue of plastic pollution by taking important steps:

Plastic Bag Ban (2017): Kenya forbade the production, distribution, and use of plastic bags and enacted one of the strongest bans on plastic bags worldwide. The severe fines or jail time that infringers risk serve as a powerful disincentive to using plastic bags.

Campaigns for Public Awareness

Consumers can learn about the advantages of switching to sustainable alternatives and using less plastic by participating in educational programs. To reach a large audience, these campaigns might make use of a variety of media platforms, such as social media, television, and local events. For instance, the "Refuse Plastic" campaign urges people to select reusable items instead of single-use plastic ones. Activities in School regarding the reduced consumption should be encouraged.

Producer Responsibility Extended (EPR)

Manufacturers are held responsible for the complete lifecycle of their products, including recycling and disposal, under EPR regulations. Producers are encouraged to design items that are less hazardous to the environment or that are easier to recycle by requiring them to manage the waste that their products generate under EPR systems. A number of nations, such as Canada and Japan, have put EPR systems in place for different kind of plastic products.

Rewards for Eco-Friendly Decisions

Financial rewards may persuade customers to embrace more environmentally friendly behaviours. Reward schemes for cutting back on plastic use, discounts for bringing reusable bags or containers, and financial aid for buying environmentally friendly goods are a few examples. For example, supermarket businesses sometimes give their patrons who bring their own reusable plastic bags.

Designing Sustainable Products

Enhancing recyclability and lowering plastic usage are two benefits of designing products with sustainability in mind. This entails minimizing packing, utilizing biodegradable materials, and designing for simple disassembly. Companies like Coca-Cola and Unilever, for instance, are producing packaging using recycled materials or biodegradable substitutes.

Adopting the Principles of the Circular Economy

Reusing, reusing, and repairing resources to extend their useful life is the main goal of the circular economy. Businesses can implement the concepts of the circular economy by creating products that are robust, recyclable, or upgradeable. The Ellen MacArthur Foundation collaborates with companies to develop sustainable solutions and supports circular economy principles.

Multiple Stakeholder Collaborations

Governments, corporations, and non-governmental organizations working together can advance sustainable practices and the decrease of plastic trash. Collaborations like the Global Partnership on Marine Litter (GPML) aim to create and execute strategies to lessen the pollution caused by plastic in the ocean.

Being a Responsible Citizen of the World

- On an individual level we should avoid using single use plastics such as drinking straw. We can always use a non- plastic straw, though these single use plastics are very conventional but they affect the planet and mainly the living beings of the oceans.
- We can always carry a cloth back while shopping and ask the shopkeeper to not put anything in plastic.
- Purchasing more bulk food than fewer packaged goods.
- Using steel utensils/containers and avoid plastic Tupperware.
- Recycling chewing gums or we can use natural chewing gums.
- Using wooden pegs after washing clothes.
- Discarding plastic waste in the proper recycling container.
- Stop using cosmetics that contains micro-plastics, purchasing biodegradable make up products or cosmetics.
- Minimise buying water sold in plastic bottles.
- Purchasing of second-hand plastic items.

Reducing the use of plastic is a difficult but necessary endeavour that calls for an all-encompassing strategy. We can address the issues raised by plastic pollution and move toward a more sustainable future by putting legislative measures into place, stimulating industrial solutions, encouraging sustainable consumer behaviour, and fostering global collaboration. We can lessen our dependency on plastic and improve the environment for now and future generations by working together and being committed.

II. **Disposal of Plastic**

The disposal of plastic management circulates around the four F's which are Reduce, Reuse Recycle and Recover. The main focus is to reduce the production of plastic and reduce the consumption both of which has been discussed earlier. Even though, the steps of production and consumption of plastic is thoroughly examined, still plastic waste are inevitably produced and this waste must be properly handled by appropriate waste management system [18]. Because it necessitates the recovery, sorting and refilling of containers, reusing packaging may be difficult and is usually reserved for expensive goods like electronics [11] car waste and can be used for energy recovery and reused as feedstock if recycling is not practical. Landfill should only be used for leftover garbage such as ash.

Good solid waste management techniques can help cut down on the amount of plastic in the environment and, in turn, stop it from breaking down into tiny pieces called microplastics [24]. For example, Taiwan has effectively reduced waste disposal rates from 0.9 to 0.48 kg per person thanks to improvements in waste management policies, such as the "Plastic Restriction Policy," which forbids the use of plastic bags and disposable plastic tableware, and the Recycling Act and the Compulsory Trash-Sorting Policy, which require waste sorting. The number of plastic bottles, plastic bags, and metal drink cans that were found on beaches significantly decreased as a result of these actions [5]. Similar results were observed in Australia, where beach litter levels were lower in towns with larger waste management budgets.

So, this portion will discuss various aspects (both the traditional and modern methods) of managing the ways for disposal of plastic waste and potential solutions to mitigate these effects. The three major categories of plastic waste are municipal solid waste, marine litter and industrial waste. All these wastes should be manged properly or they will lead to various pollution in the environment which will affect almost all life forms.

6 Traditional Method of Plastic Waste Disposal

6.1 Landfilling:

Landfills and the Disposal of Plastic Waste

Plastics usually wind up in landfills after being thrown into various rubbish containers. Before being buried beneath the surface of the earth, throwaway plastic waste is placed in landfills. Care must be taken when using this kind of garbage disposal to prevent unfavourable outcomes that could result from insufficient processing, such as soil damage and groundwater contamination [25]. Establishing a secure disposal area with the least amount of negative environmental impact is the main goal of landfill management, safeguarding air quality and aquatic habitats. Significant work is required to achieve these aims, such as excavating trenches or deep pits to deposit garbage and let it break down over time. This breakdown process might take years to finish and is frequently quite slow [10].

Every molecule in the garbage experiences biodegradation as the organic compounds break down. Because of the special biochemical characteristics of plastics, the decomposition of garbage made of long-chain polymers, such bags, can take anywhere from ten to one hundred years. Factors such as sunlight, wind, and general climate have a major effect on the rate of degradation as well [26]. Reuse and recycling should take precedence over landfill dumping because of the long breakdown durations.

Notwithstanding the difficulties, landfills can be used as a source of energy since they produce methane and carbon dioxide as organic matter breaks down. They are essential for keeping cities clean, promoting hygienic practices, and separating garbage from harmful materials. Landfills are also an affordable way to dispose of plastic trash. But there are some significant disadvantages to this disposal technique as well, such as how it contributes to climate change since methane, a highly flammable gas, is released.

6.2 Incineration

Carbon dioxide and water vapor are released during the process of burning waste materials in the presence of oxygen. Because of their chemical makeup, some plastic waste types are not appropriate for incineration; others may even be explosive. Plastics must be carefully chosen and sorted in order to prevent hazardous situations during the burning process. The capacity of incineration to significantly reduce trash volume is one of its main benefits. In addition to reducing the amount of physical trash produced, this process produces energy—heat and electricity that is necessary for both household and commercial use. Incineration has proven especially successful in Europe, where it helps with trash reduction and the production of renewable energy. Incineration has drawbacks despite its advantages, which include lessening the need for landfills, lowering environmental pollution, and producing energy. In comparison to alternative waste management techniques, the establishment expenses of incinerator facilities are higher. Furthermore, the procedure may cause dangerous air pollutants to be released, endangering both human and wildlife health. It's also necessary to handle the incinerated ash residue carefully to avoid contaminating the environment.

In conclusion, incineration is essential to waste management since it reduces volume of waste and produces energy, but its effects on the environment and human health must be carefully considered before it is used. Incineration is used to manage about 26% of municipal garbage in the European Union.

6.3 *Open Dumping*

Open dumping is an unregulated form of disposing of rubbish that is usually done in public areas, particularly on the outskirts of cities. Because there is insufficient infrastructure for effective garbage management, this practice is widespread in many underdeveloped nations. Open dumping causes serious environmental damage, including greenhouse gas emissions, the spread of pests that spread illness, and contamination of the land and water. It also presents hazards to people's safety and health.

7 New Techniques for Disposal of Plastic Waste

7.1 *Pyrolysis*

The process of recovering crude petrochemicals and obtaining hydrocarbons by converting gasses and fatty oils is known as pyrolysis. It is even utilized to produce sustainable energy from plastic waste and recover crude petrochemicals [20]. The three primary categories of pyrolysis process are based on the heat energy required to break plastic linkages. There are media that is temperature-based for high, medium, and low temperatures. The range of temperatures required to demolish the plastic structure determines what constitutes medium and high temperatures [20]. About 80% of the plastics used in everyday human necessities are classified as thermosets or thermoplastics.

This is predicated on their adaptability to change and their facile molecular reformation after heat treatment. The byproducts of pyrolysis of plastics, contingent upon several parameters such as reactor type, residence duration, composition of plastics, condensation configuration, feeding configuration and the applied temperature [12]. Through catalytic chemical treatment or heat effects, the crude oil derivative plastics are transformed to its monomers and other useful components, such as additives and plasters, which are classified into the first derivative of petroleum refining cuts and petrochemicals [21]. In the waste management sector, these chemical treatment guidelines were developed to ensure appropriate handling and collection of plastic wastes. This scientific approach, while rather expensive up front, is an effective way to manage waste.

7.2 *Bioremediation*

Utilizing microorganisms including bacteria, fungi, algae, and plants, bioremediation breaks down trash. For this technological process to efficiently break down natural and manmade chemicals, ideal circumstances including the right nutrients, enzymes, temperature, and pressure are needed. Plastic polymers are changed into less toxic byproducts by the process of bioremediation, which makes use of enzymes like oxidoreductase and peroxidases. Although bioremediation presents a sustainable substitute for incineration and landfills, it frequently necessitates a substantial financial outlay to facilitate the efficient breakdown of synthetic substances.

7.3 Chemical Recycling

Chemical recycling, sometimes referred to as feedstock recycling, is the process of disassembling plastics into their component chemicals. After being refined, these building blocks can be utilized to create valuable compounds or new plastics. Compared to mechanical recycling, this process enables recycling of a wider variety of plastic kinds and preserves the quality of recycled materials. Chemical recycling is still in its infancy, but it has the potential to overcome the drawbacks of conventional recycling techniques.

7.4 Biodegradation

Plastic garbage is broken down by biological processes through composting and biodegradation. Under some circumstances, biodegradable plastics composed of natural substances like cornstarch and sugarcane can break down. Biodegradable plastics can be processed at composting facilities to create organic matter that improves soil quality. Nevertheless, there are obstacles to the broad use of biodegradable plastics. These include the requirement for particular circumstances in order to accomplish total decomposition and the possibility of contaminating traditional plastic waste streams.

All these techniques are used for disposal of plastic, improper disposal leads to release of toxins into the environment that can cause human health risk and deteriorates the environment. Most of which we have discussed previously. Once these plastics enters the food chain, it is irreversible to eliminate its deadly consequences so it is better to take all precaution while disposing off the plastic and improve the disposal strategies to better techniques with time. Of all the strategies international cooperation and policy developing is must as it is a global issue.

Governments ought to collaborate to create uniform policies and procedures for handling plastic garbage. International accords, like the Basel Convention, can support environmentally responsible disposal methods and assist regulate the transboundary movement of plastic garbage. Working together can also make it easier to share innovations and best practices.

8 Conclusion

Getting rid of plastic waste is a difficult task that calls for a multifaceted strategy. There are serious concerns to the environment and human health associated with traditional disposal techniques like landfilling and incineration. New techniques including biodegradation, pyrolysis, and recycling are viable alternatives but need more research and funding. A mix of improved recycling infrastructure, international cooperation, policy development, source reduction, and research and innovation are needed for effective management of plastic trash. By using these tactics, we can lessen the negative effects that plastic trash has on the environment and human health and work toward a more sustainable future.

9 Prospects for the Future

In order to fully solve the issue, future efforts should concentrate on creating novel solutions for the disposal of plastic trash, raising public awareness and educating the public, and bolstering international cooperation. To establish a sustainable and circular economy that places a high priority on the reduction, reuse, and recycling of plastic materials, governments, businesses, and individuals must collaborate.

References

1. Abbasi S, Keshavarzi B, Moore F, Turner A, Kelly FJ, Dominguez AO, Jaafarzadeh N (2019) Distribution and potential health impacts of microplastics and microrubbers in air and street dusts from Asaluyeh County, Iran. Environ Pollut 244:153–164
2. Al-Salem SM, Lettieri P, Baeyens J (2009) Recycling and recovery routes of plastic solid waste (PSW): a review. Waste Manag 29:2625–2643
3. Anderson AG, Jane Grose, Sabine Pahl, Richard C. Thompson, Wyles KJ (2016) Microplastics in personal care products: exploring perceptions of environmentalists, beauticians and students. Mar Pollut Bull 113(1–2):454–460
4. Beitzen-Heineke EF, Balta-Ozkan N, Reefke H (2017) The prospects of zero-packaging grocery stores to improve the social and environmental impacts of the food supply chain. J Clean Prod 140:1528–1541
5. Brennholt N; Heß M, Reifferscheid G (2018) Freshwater microplastics. Springer International Publishing, Cham, Switzerland, pp 239–272
6. Chang and Michelle (2015) Reducing microplastics from facial exfoliating cleansers in wastewater through treatment versus consumer product decisions. Mar Pollut Bull 101(1):330–333
7. Evode N, Qamar SA, Bilal M, Barceló D, Iqbal HMN (2021) Plastic waste and its management strategies for environmental sustainability. Case Stud Chem Environ Eng 4:100142
8. Kumari R, Tiwari R, Akhtar R, Gupta SK (2023) Harmful effects of plastics on human health and the environment: a review. GPH-Int J Soc Sci Hum Res 6(11):248–255
9. Li B, Ding Y, Cheng X, Sheng D, Xu, Zheng, Rong Q, Wu, Yulong, Zhao H, Ji X, Zhang Y (2020) Polyethylene microplastics affect the distribution of gut microbiota and inflammation development in mice. Chemosphere 244:125492

10. Liang Y, Tan Q, Song Q, Li J (2021) An analysis of the plastic waste trade and management in Asia. Waste Manage 119:242–253
11. Liu Z, Adams M, Walker TR (2018) Are exports of recyclables from developed to developing countries waste pollution transfer or part of the global circular economy? Resour Conserv Recycl 136:22–23
12. Miandad R, Barakat MA, Aburiazaiza AS, Rehan M, Nizami AS (2016) Catalytic pyrolysis of plastic waste: a review. Process Saf Environ Protect 102:822–838
13. Ragusa A, Svelato A, Santacroce C, Catalano P, Notarstefano V, Carnevali O, Papa F et al (2021) Plasticenta: First evidence of microplastics in human placenta. Environ Int 146:106274
14. Rayner JL, Wood C, Fenton SE (2004) Exposure parameters necessary for delayed puberty and mammary gland development in Long-Evans rats exposed in utero to atrazine. Toxicol Appl Pharmacol 195(1):23–34
15. Reddy MS, Srinivasulu Reddy P, Venkata Subbaiah G, Venkata Subbaiah H (2014) Effect of plastic pollution on environment. J Chem Pharm Sci 14:28–29
16. Rudel RA, Camann DE, Spengler JD, Korn LR, Brody JG (2003) Phthalates, alkylphenols, pesticides, polybrominated diphenyl ethers, and other endocrine-disrupting compounds in indoor air and dust. Environ Sci Technol 37(20):4543–4553
17. Rudel RA, Brody JG, Spengler JD, Vallarino J, Geno PW, Sun G, Yau A (2001) Identification of selected hormonally active agents and animal mammary carcinogens in commercial and residential air and dust samples. J Air Waste Manag Assoc 51(4):499–513
18. Schneider DR, Ragossnig A (2015) Recycling and incineration, contradiction or coexistence? Waste Manag Res 33(8):693–695
19. Schönfelder G, Wittfoht W, Hopp H, Talsness CE, Paul M, Chahoud I (2002) Parent bisphenol A accumulation in the human maternal-fetal-placental unit. Environ Health Perspect 110(11):A703–A707
20. Sharuddin SDA, Abnisa F, Daud WMAW, Aroua MK (2016) A review on pyrolysis of plastic wastes. Energy Convers Manag 115:308–326
21. Shent H, Pugh RJ, Forssberg E (1999) A review of plastics waste recycling and the flotation of plastics. Resour Conserv Recycl 25(2):85–109
22. Wagner M, Oehlmann J (2009) Endocrine disruptors in bottled mineral water: total estrogenic burden and migration from plastic bottles. Environ Sci Pollut Res 16(3):278–286
23. Warner M, Eskenazi B, Mocarelli P, Gerthoux PM, Samuels S, Needham L, Patterson D, Brambilla P (2002) Serum dioxin concentrations and breast cancer risk in the Seveso Women's Health Study. Environ Health Perspect 110(7):625–628 (2002)
24. Wu W-M, Yang J, Criddle CS (2017) Microplastics pollution and reduction strategies. Front Environ Sci Eng 11:1–4
25. Zheng Y, Yanful EK, Bassi AS (2005) A review of plastic waste biodegradation. Crit Rev Biotechnol 25(4):243–250
26. Zhou C, Fang W, Xu, Wanying, Cao A, Wang R (2014) Characteristics and the recovery potential of plastic wastes obtained from landfill mining. J Clean Prod 80:80–86

Sustainability Practices in Emerging Markets: Overview and Trends

Oluwasikemi Janet Owolabi

Abstract This study reviewed sustainability practices of firms in the emerging markets as aligning with the recent development of sustainability practices and reporting is crucial for the firms. Content analysis was used to quantify the practices of sustainability (environmentally, economically, socially, and governance-wise) of listed deposit money banks in Nigeria against the GRI sustainability guidelines. From the trend analysis and descriptive statistics carried out the study found that the practice of sustainability by these firms had been dwindling ranging from non-compliance to full compliance with the GRI requirements. Thus, the study recommends that the policy makers and regulators in the emerging nations should intensify efforts to encourage the continuous compliance of these firms with sustainability requirements.

Keywords Economic sustainability · Emerging markets · Environmental sustainability · Global reporting initiatives · Governance sustainability · Social sustainability · Sustainability practices

1 Introduction

Globally, sustainability performance has been on the increase, becoming key business issues that are affected by various market forces. Sassen et al. [31] argue that companies actively practicing and transparently disclosing their environmental, economic, social, and governance risks tend to enjoy a stronger position for generating shareholder wealth.

For more than two decades, organizations have been disclosing non-financial information to foster transparency, communication, and accountability with their stakeholders [32]. Firms face the challenge of not merely disclosing financial data but also integrating non-financial information to cultivate stakeholders' trust [27]. Consequently, the annual reports of many companies now encompass economic, social, environmental, and governance data, furnishing stakeholders with comprehensive

O. J. Owolabi (✉)
Accounting Department, Pan-Atlantic University, Lagos State, Ibeju-Lekki, Lekki, Nigeria
e-mail: oowolabi@pau.edu.ng

J. A. Parray et al. (eds.), *Sustainable Urban Environment and Waste Management*, Urban Sustainability, https://doi.org/10.1007/978-981-96-1140-9_10

information for decision-making. This practice is commonly known as sustainability reporting [28].

Sustainability entails meeting current needs in a manner that preserves the ability to meet the needs of future generations. Companies are actively pursuing social and environmental responsibility by mitigating the diverse impacts of their operations, including social issues, pollution, global warming, and upholding principles of good corporate governance [26].

Sustainability entails meeting current needs in a manner that preserves the ability to meet the needs of future generations. Companies are actively pursuing social and environmental responsibility by mitigating the diverse impacts of their operations, including social issues, pollution, global warming, and upholding principles of good corporate governance [26].

In many countries, including Nigeria, the adoption of sustainability reporting by firms remains voluntary, despite its critical importance. Consequently, the commitment to sustainability practices among firms in these nations has been waning. The inaugural sustainability reporting took place in 1999, aligning with the Global Reporting Initiative (GRI) framework [21]. Notably, the forthcoming efforts of the Global Reporting Initiative are primarily directed towards developing countries, recognizing their urgent requirement for sustainability reporting. These nations confront severe economic and environmental challenges, exacerbated by rapid population growth [13].

According to Lizinska and Czapiewski [20], the long-term sustainable wealth of companies is strongly connected with their activities. Thus, corporate sustainability practices are expected to enhance the quality of earnings and long-term value of firms. Sustainability practices and reporting is a key element of corporate financial management and it‘s crucial to debtholders, investors and other stakeholders [19]. Alsayegh et al. [2] highlight that organizations are increasingly acknowledging the advantages of incorporating sustainability strategies to advance their future objectives. Such benefits include enjoying legitimacy of firms, competitive advantage, enhanced performance, product differentiation, access to funds at lower costs, enhanced going concern, and better stakeholder relationship [34].

2 Theoretical Underpinnings and Review of Literature

Overtime, scholars have considered various theories that portray the essence of sustainability practices by firms which include agency theory, signaling theory, stakeholder theory, and legitimacy theory. This study concentrated on exploring signaling theory and stakeholder theory to buttress the importance of sustainability practices and reporting by organisations.

Signaling Theory developed by Michael Spencer in 1973, assumes that information is not usually available at the same time to all parties which resulted to information asymmetry. Signaling theory elucidates the strategic actions undertaken by firms to mitigate information asymmetries between themselves and their stakeholders. This imbalance is addressed through proactive information disclosure. Firms that exhibit a commitment to enhanced information transparency signal to their stakeholders, thereby differentiating high-performing companies from their counterparts. In alignment with this theory, an uptick in voluntary disclosure not only fosters public loyalty but also has the potential to stimulate increased demand for a company's shares, products, and services. This, in turn, contributes to the creation of sustainable value, reduced cost of capital and ensures the long-term survival of the company.

Stakeholder Theory, pioneered by Edward Freeman in 1984, asserts that an organization should proactively engage in meeting the diverse needs of various constituents who wield influence over the firm's outcomes. It perceives organizations as intricate systems that must consider not only the interests of shareholders but also those of other groups within the operational environment. This perspective stands in contrast to agency theory, which interprets organizations as structures defined by the relationships between shareholders and management [22]. Hence, organizations should prioritize shareholders' returns and also those of the other class of stakeholders by practicing and reporting sustainability in accordance with the Global Reporting Initiative (GRI) sustainability requirements.

The term sustainability has been used in several dimensions. Often, sustainability is used as sustainability development [11]. At the development and implementation stages of sustainability framework, several international conferences were held for the purpose of reviewing the progresses made. The term corporate social responsibility (CSR) and sustainability are usually used interchangeably. However, they differ in relation to time as sustainability ensures that corporate decisions and activities are short and long term concerned [5].

In 2023, the International Sustainability Standards Board (ISSB) introduced IFRS S1 and IFRS S2, outlining essential guidelines for companies to disclose their sustainability-related data. IFRS S1, denoting General Requirements for Disclosure of Sustainability-related Financial Information, and IFRS S2, representing Climate Related Disclosures, aim to standardize reporting practices. These regulations are slated to be implemented for reporting periods starting from January 2024 onwards. Figure 1 shows the roadmap for the adoption of IFRS S1 and S2 in Nigeria.

Miralles-Quiros et al. [23] refers to sustainability practices as environmental and social activities of firms to present themselves as responsible firms to their stakeholders. Also, Gonçalves et al. [15] opined that sustainability practices are firm's engagement to create and build moral capital with stakeholders which could be used as risk management tool. Similarly, Salvi et al. [30] provided a comprehensive definition of sustainability practices, describing them as the actions undertaken by organizations to uphold sustainability across four key dimensions: economic, environmental, social, and corporate governance. This definition underscores the holistic approach that organizations should adopt to address sustainability challenges and ensure long-term viability and responsible management. World Council for Economic Development (WCED) defined sustainability practices as "meeting the needs of the present without compromising the ability to meet the needs of future generations".

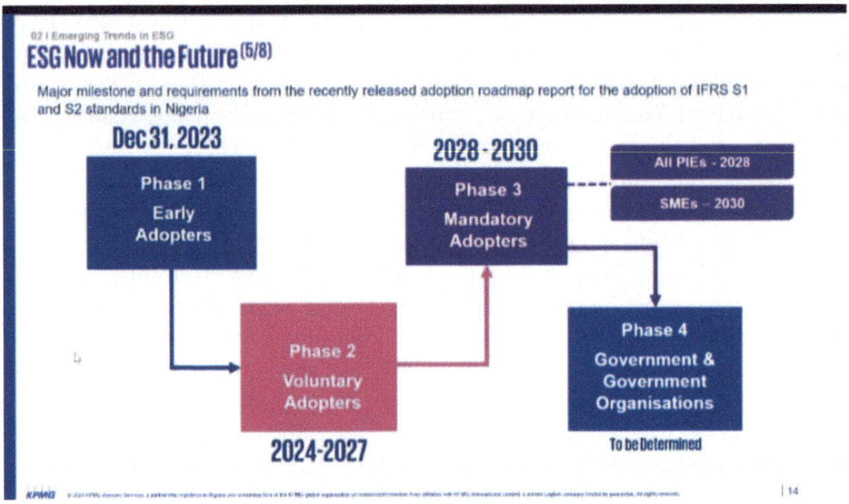

Fig. 1 Roadmap for adoption of IFRS S1 and S2 in Nigeria. *Source* Adapted from KPMG [17]

According to Anekwe et al. [3] sustainability is an approach used by organizations to derive true values for themselves as well as for the resources utilized in generating such values. Similarly, Dibia and Nwaigwe [12] opined that sustainability practices are firms' activities that create a balance between human use of the regenerative capacity of the environment and the allocation of the natural capital to enhance development. Sustainability practices have also been defined as firms' activities that create innovation and competitive advantage [1]. Also, Buallay [8] defines sustainability practices as activities of firms towards meeting the economic, environmental and social needs of present and future generations through corporate governance. Therefore, the study defines sustainability practices as practices that firms engage in to ensure that they firms meet its current needs and ensure that its going concern is not jeopardized. It encompasses economic sustainability, environmental sustainability, social sustainability, and governance sustainability. Over the years, the term sustainability has been broadened to encompass governance, social, environmental and economic dimensions. This is to indicate that firms are not operating with a single goal of creating economic values rather to ensure a sustainable future [11, 14].

The economic dimension has been defined as activities of firms that focus on maintaining market presence through financial performance improvement [6]. Hover, Dembo [11] suggested that the focus of economic sustainability practices is to promote firms' long-run economic performance. According to GRI [14], "economic sustainability practices are activities of organisations that aimed at improving the economic condition of the stakeholders at the local, national and global level". Also, NSE [25] provided that the contribution of the firm to the larger economy is the focus of users of sustainability reports rather than the financial performance which is usually disclosed in financial statement. The economic sustainability practices has four aspects of practice and each has its indicators. The aspects encompass economic performance, market presence, indirect economic impact, and procurement practices [14].

Environmental sustainability practices has been defined has activities of firms focused on achieving financial performance without jeopardizing the ability of internal and external resources in providing long term growth [16]. According to Dibia and Nwaigwe [12] environmental sustainability practices are firms' practices focused on minimizing the negative impact of its activities on ecosystem. Also, Taiwo et al. [33] defined environmental sustainability practices as organizational practices directed towards reducing the negative impact of its activities on living and non-living natural system. Human activities usually affect the environment adversely by causing climate change, global warming and degradations, as such, firms should be center to providing solution [4]. According to GRI [14], environmental sustainability practices focuses on "effect of inputs (such as energy and water), outputs (such as emissions, effluents and waste), environmental compliance and expenditures, materials, biodiversity, transport, and product and service". Also, Ramirez et al. [29] refers to environmental sustainability practices as actions of firms geared towards environmental responsibility as related to innovation, use of resources and emission. Environmental sustainability practices have eleven aspects of Product/ service responsibility, water, energy, effluents and waste, compliance, environmental grievance mechanism, materials, emissions, transport, and overall [14].

According to Buallay [8], social sustainability practices consider social issues and adverse effect of social activities on the society. Also, Taiwo et al. [33], social sustainability practices focus on maintaining positive social values for the society. Likewise, institutional theory provides that stakeholders' pressure is the driver of social sustainability practices. Also, Ramirez et al. [29] defined social sustainability practices as actions of firms that demonstrate their commitment to the environment where they operate and to all areas of their supply chain. Thus, social sustainability practices emphasis customer's satisfaction, employees and societal wellbeing. Social sustainability practices have four aspects of Labor practices and decent work, human rights, society, and product responsibility [14].

Governance sustainability practices involve principles that assist firms to enhance transparency and prevent conflict of interest [8]. According to Buallay et al. [7], governance sustainability practices ensure compliance with rules, regulations and laws relating to social, economic, and environmental practices. It encompasses firms' practices aimed at safeguarding the interests of stakeholders [33]. Also, Korolo and Korolo [18] defined governance sustainability dimension as the practices that align the goals of economic, social and environmental practices. Ramirez et al. [29] refers to governance sustainability practices as systems and processes used by an entity to ensure that the actions of its management and workers are geared towards protecting the interest of the stakeholders. Corporate governance dimension is important to stakeholders because it is the only medium through which management actions could be monitored. Governance dimension has six aspects of: Anti-corruption, governance structure and composition, role of highest governance body in sustainability reporting, role of highest governance body in evaluating economic, environmental and social practices, role of highest governance body in risk management, and role

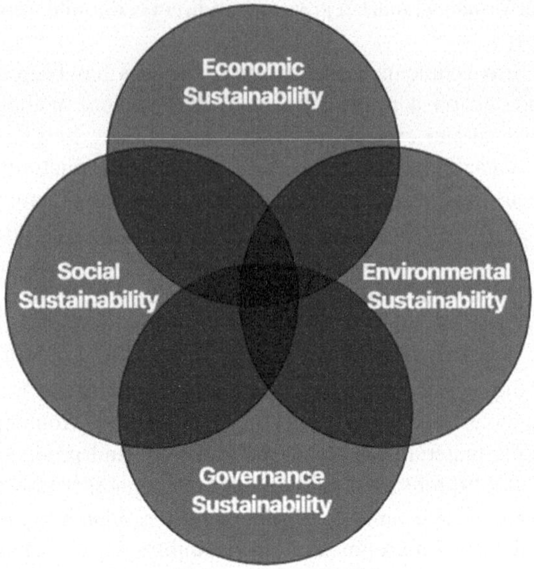

Fig. 2 The four dimensions of sustainability practices. *Source* Author (2024)

of highest governance body in seeking value and strategy [14]. Figure 2 illustrates the four dimensions of sustainability practices.

2.1 Global Reporting Initiatives (GRI) Sustainability Guidelines and Nigerian Sustainable Banking Principles

Globally, the GRI guidelines have been acknowledged as the predominant standard for sustainability practices and reporting. The Coalition for Environmentally Responsible Economies (CERES) in the United States and the United Nations Environment Programme (UNEP) jointly issued the GRI in 1997 [35]. To encompass a holistic sustainability dimension of economic dimension, environmental dimension, social dimension and governance dimension, the framework was reviewed in 1998. The GRI standard was developed in four versions with the 4th version issued in 2013. The GRI framework is established to be adopted in enhancing transparency by firms regardless of their location, size and sector [24].

Nigerian sustainable banking principles was issued by the Central Bank of Nigeria in 2012 with the aim of enhancing the long-term financial services of the institutions while ensuring the institutions are environmentally and socially responsible [9]. The Nigerian sustainable banking principles are developed to address the unique local environment challenges and priorities for financial institutions in Nigeria. The principles encompass environmental and social risk management, minimizing environmental and social footprint, upholding human rights, promoting women's economic empowerment, fostering financial inclusion, strengthening environmental and social governance, enhancing capacity building, fostering collaborative partnerships, and ensuring transparent reporting [9].

However, for the purpose of global comparability since Nigerian banks also seek to attract foreign investors and other stakeholders, the GRI sustainability standards was used for the study. Also, GRI provides more comprehensive framework for firms to practice and report sustainability practices which enhances transparency and comparability.

3 Methodology

By making use of *ex-post facto* research design the study obtained secondary data on sustainability practices of listed deposit money banks (DMBs) as at 31st December, 2022 in Nigeria from their annual reports and sustainability reports for the period of 2012–2022. This is to enable the review of sustainability practices (economic sustainability, environmental sustainability, social sustainability and governance sustainability) of the listed DMBs for the period covered. The approaches of Ching et al. [10] and Taiwo et al. [33] were used through content analysis to quantify the sustainability practices against Global Reporting Initiative (GRI) sustainability guidelines.

4 Trend Analysis of Practices of Sustainability Dimensions by Listed Deposit Money Banks in Nigeria

The graphical trend analysis shown below are those of the dimensions of the sustainability (environmental sustainability, economic sustainability, social sustainability and governance sustainability) practices by the listed deposit money banks in Nigeria. The figures below indicates that the practices of sustainability by these banks have been dwindling ranging from non-compliance to full compliance by the banks with the GRI sustainability requirements this further indicate that these banks are used to the traditional ways of carrying out their operations and trying to see how to adjust to the recent development of sustainability practices.

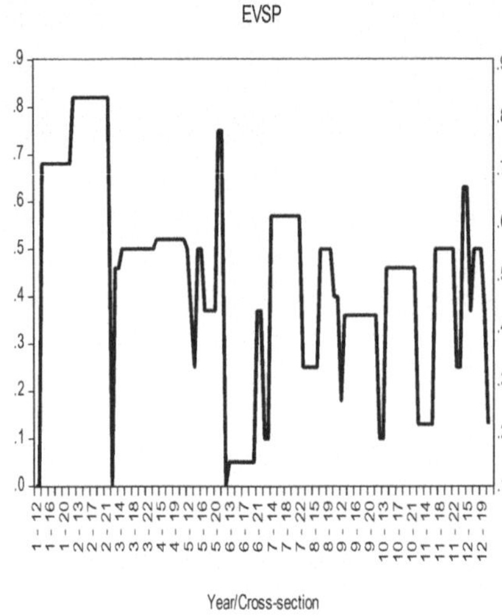
EVSP
Year/Cross-section

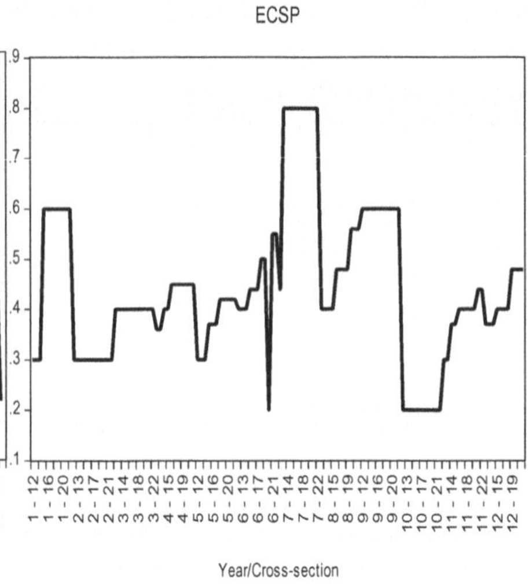
ECSP
Year/Cross-section

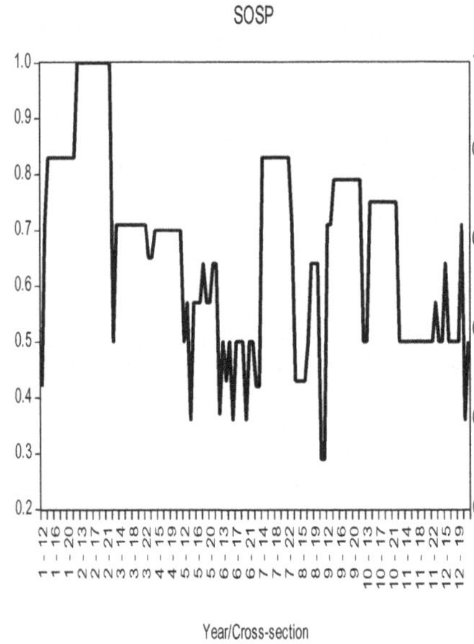
SOSP
Year/Cross-section

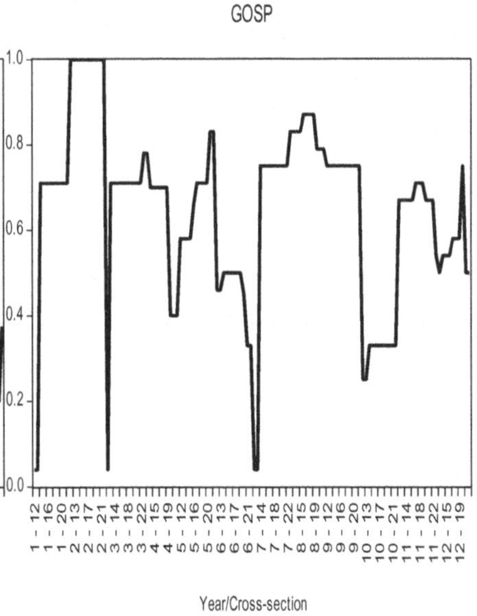
GOSP
Year/Cross-section

5 Descriptive Statistics

This section is used to describe the dimensions of sustainability practices through mean, maximum, minimum and standard deviation. It summarizes the dimensions of sustainability practices (Environmental sustainability practices (EVSP), Economic sustainability practices (ECSP), Social sustainability practices (SOSP), and Governance sustainability practices (GOSP)).

Interpretation

From Table 1, **EVSP**: The mean value of environmental sustainability practices is 0.4309 and the standard deviation is 0.2288 the mean value of 43% is relatively low which means that on average the banks practices environmental sustainability to the tune of 43% in accordance with GRI sustainability guidelines. The standard deviation of 22% indicates that the level of variation in practices of environmental sustainability is low. The minimum value of 0.05 indicates that some of the banks lowly practiced environmental sustainability while the maximum value of 0.82 indicates that some of the banks highly practiced environmental sustainability.

ECSP: Economic sustainability practices has a mean value of 0.4431 and standard deviation of 0.1518. The mean value of 44% suggests that averagely, the banks practiced economic sustainability to the tune of 44% which is relatively low as it is below average. The standards deviation of 15% suggests that the economic sustainability practices of the banks are close to the mean value. The minimum value of 0.2 and maximum value of 0.8 indicate that some banks lowly [practiced economic sustainability while other highly practiced it in line with GRI sustainability guidelines.

SOSP: The mean value of 0.6539 and standard deviation of 0.1927 suggest that on average, the deposit money banks highly practiced social sustainability to the tune of 65% with a low dispersion in their social sustainability practices to the tune of 19%. Also, the minimum value of 0.2 and maximum value of 1 indicate that some of the banks lowly practiced social sustainability while other practiced it completely as required by GRI sustainability guidelines.

GOSP: Governance sustainability practices have the mean of 0.6472 and standard deviation of 0.2147. The mean of 64% is moderately high and indicates that the

Table 1 Descriptive statistics result

Variable	Mean	Std. Dev	Minimum	Maximum	Obs
EVSP	0.4309	0.2288	0.05	0.82	132
ECSP	0.4431	0.1518	0.2	0.8	132
SOSP	0.6539	0.1927	0.2	1	132
GOSP	0.6472	0.2147	0.04	1	132

Source Researcher's computation (2024)

deposit money banks on average practiced governance sustainability practices to the tune of 64%. Also, a standard deviation of 21% means that the governance sustainability practices of the banks are not to far from the mean value. The minimum value of 0.04 and maximum value of 1 suggest that some of the banks lowly practiced governance sustainability while others completely practiced it.

6 Conclusion and Recommendation

The study had reviewed the sustainability practices in emerging market by considering the practices of listed DMBs in Nigeria. Based on the trend analysis and descriptive statistics as shown above, the study concludes that firms in emerging markets are yet to fully comply with the sustainability requirements of GRI which could deny them the various associated benefits of practicing sustainability. Therefore, the study recommends that policy markets and regulators in the emerging markets should continually encourage their listed firms to be intentional in aligning with the new development of sustainability practices and reporting to enjoy the advantage of improved reputation, assess to funds at lower costs, attract more investors, competitive advantage and enhanced going concern.

References

1. Ahmed A, Ahmed D (2018) The impact of social, environmental and corporate governance disclosures on firm value Evidence from Egypt. J Account Emerg Econ 8(4):442–458
2. Alsayegh MF, Abdul Rahman R, Homayoun S (2020) Corporate economic, environmental, and social sustainability performance transformation through ESG disclosure. Sustainability 12(9):3910
3. Anekwe RI, Ndubusi-Okolo P, Uzoezie C (2019) Sustainability in the Nigerian business environment: problems and prospects. Int J Acad Manag Sci Res 3(3):72–80
4. Asuquo AI, Dada ET, Onyeogaziri UR (2018) The effect of sustainability reporting on corporate performance of selected quoted brewery firms in Niger Int J Bus Law Res 6(3):1–10
5. Backstrom S, Kalsson J (2015) Corporate sustainability and financial performance: The influence of board diversity in a Swedish context. Master's thesis, Uppsala University. https://pdfs.semanticscholar.org/520d/2c70f03e2f91ad50731fcb3985dffc982a8c.pdf?ga=2.25801977.555165174.1570734734-1198418072.1569176799
6. Baumgartner RJ, Ebner D (2010) Corporate sustainability strategies: sustainability profiles and maturity levels. Sustain Dev 18(2):76–89
7. Buallay A, Hamdan A, Zureigat Q (2017) Corporate governance and firm performance: evidence from Saudi Arabia. Australas Account Bus Financ J 11(1):78–98
8. Buallay AM (2020) The level of sustainability reporting and its impact on firm performance: the moderating role of a country's sustainability reporting law. A PhD thesis of Brunel University, London
9. Central Bank of Nigeria (CBN) (2012) Nigerian sustainable banking principles. Retrieved from https://www.cbn.gov.ng/out/2012/ccd/circular-nsbp.pdf
10. Ching HY, Gerab F, Toste TH (2017) The quality of sustainability reports and corporate financial performance: evidence from Brazilian listed companies. J SAGE Open 1–9

11. Dembo AM (2017) The impact of sustainability practices on the financial performance: evidence from listed oil and gas companies in Nigeria. In: Dimensional corporate governance. Springer, Cham, pp 215–233
12. Dibia NO, Nwaigwe NG (2018) Sustainable development practices and corporate financial performance: a survey of selected quoted companies in Nigeria. Asian J Econ Bus Account 1–13
13. Global Reporting Initiatives (GRI) G3 (2011) A new phase: the growth of sustainability reporting, 3
14. Global Reporting Initiative (GRI) G4 (2015) Sustainability reporting guideline
15. Gonçalves TC, Dias J, Barros V (2022) Sustainability performance and the cost of capital. Int J Financ Stud 10(3):63
16. Hoepner A, Oikonomou I, Scholtens B, Schröder M (2016) The effects of corporate and country sustainability characteristics on the cost of debt: an international investigation. J Bus Financ Acc 43(1–2):158–190
17. KPMG (2023) Roadmap for IFRS S1 and S2 adoption in Nigeria. Retrieved from https://kpmg.com/ng/en/home/insights/2023/07/issb-standards-are-now-live.html
18. Korolo EO, Korolo AS (2023) Corporate sustainability reporting and financial performance of deposit money banks in Nigeria. FUDMA J Account Financ Res [FUJAFR] 1(3):75–86
19. Kumar K (2022) Emerging phenomenon of corporate sustainability reporting: evidence from top 100 NSE listed companies in India. J Public Aff 22(1):e2368
20. Lizińska J, Czapiewski L (2018) Towards economic corporate sustainability in reporting: what does earnings management around equity offerings mean for long-term performance? Sustainability 10(12):4349
21. Loh L, Thomas T, Wang Y (2017) Sustainability reporting and firm value: Evidence from Singapore listed companies. J Sustain 9(5):1–12
22. Mihai C, Leontina P, Mihai-Bogdan A, Iuliana EG (2019) Is sustainability reporting a business strategy for firm's growth? empirical study on the Romanian capital market. J Sustain 11(658):1–21
23. Miralles-Quiros MDM, Miralles-Quiros JL, Arraiano IG (2017) Are firms that contribute to sustainable development valued by investors? Corp Soc Responsib Environ Manag 24(1):71–84
24. Najul LS (2016) Corporate sustainability reporting practices in India: myth or reality? Soc Responsib J 12(4):1–29
25. Nigerian Exchange Group (NSE) (2016) Sustainability disclosure guideline. Retrieved on 30 May 2019. http://www.nse.com.ng/regulationsite/IssuersRules/Sustainability%20Disclosure%20Guidelines.pdf
26. Niklander E (2020) The payoff of doing good-the impact of ESG criteria on firms' cost of debt capital. Unpublished M Sc, Thesis, School of Accounting and Finance, University of Vaasa, Finland
27. Nnamani JN, Onyekwelu UL, Ugwu OK (2017) Effect of sustainability accounting and reporting on financial performance of firms in Nigeria brewery sector. Eur J Bus Innov Res 5(1):1–15
28. Ould DEN (2020) Environmental, social, and governance disclosure, ownership structure and cost of capital: evidence from the UAE. Sustainability 12(18):7706
29. Ramirez AG, Monsalve J, González-Ruiz JD, Almonacid P, Peña A (2022) Relationship between the cost of capital and environmental, social, and governance scores: evidence from Latin America. Sustainability 14(9):5012
30. Salvi A, Petruzzella F, Giakoumelou A (2018) Does sustainability foster the cost of equity reduction? the relationship between corporate social responsibility (CSR) and riskiness worldwide. Afr J Bus Manage 12(12):381–395
31. Sassen R, Hinze AK, Hardeck I (2016) Impact of ESG factors on firm risk in Europe. J Bus Econ 86(8):867–904. https://doi.org/10.1007/s11573-016-0819-3
32. Schoenmaker D, Gianfrate G, Wasama S (2018) Cost of capital and sustainability: a literature review. Erasmus platform for sustainable value creation. https://www.coursehero.com/file/45416282/11-04-Cost-ofCapitalpdf/. Accessed 2 June 2022

33. Taiwo OJ, Owolabi BA, Adedokun Y, Ogundajo G (2021) Sustainability reporting and market value growth of quoted companies in Nigeria. J Financ Report Account 19(1):100–120
34. Tang M, Walsh G, Lerner D, Fitza MA, Li Q (2018) Green innovation, managerial concern and firm performance: an empirical study. Bus Strateg Environ 27(1):39–51
35. Yu M, Zhao R (2015) Sustainability and firm valuation: an international investigation. Int J Account Inf Manag 23(3):289–307

Solid Waste Treatment Technologies: Urban Sustainability Challenges

Plabani Roy, Gulab Khan Rohela, Shivam Bhardwaj, S. Syam, Basanagouda Gonal, Pawan Saini, Priyanka Saha, and Omais Bin Ayoub

Abstract The use of solid waste treatment technology is essential for tackling issues related to urban sustainability, which are becoming more and more crucial due to the world's fast urbanization. Urban areas are hubs of economic activity and population growth, resulting in escalating levels of solid waste generation including household waste, commercial waste, construction and demolition debris, and hazardous waste. Each of these waste streams requires tailored treatment approaches to ensure proper handling and disposal, posing logistical and operational challenges for urban waste management systems. As a result, there is a pressing need for innovative solid waste treatment technologies that can efficiently process and dispose of urban waste while minimizing negative repercussions. Conventional waste management practices, such as landfilling and open dumping, are no longer viable options due to limited land availability and adverse environmental impacts. New approaches may include novel ideas for by-product utilization from urban wastes. While some technologies, such as incineration and waste-to-energy, offer opportunities for energy recovery and resource utilization, they also emit greenhouse gases and other pollutants that can contribute to air and water pollution. Balancing the environmental benefits and drawbacks of different treatment options is essential for promoting sustainable urban waste management practices. Achieving urban sustainability requires holistic approaches that integrate technological innovation with social engagement, policy interventions, and institutional capacity building. Furthermore, social and economic factors influence the adoption and effectiveness of solid waste treatment technologies in urban areas. In this chapter we highlight concerned efforts across multiple fronts, including

P. Roy (✉) · G. K. Rohela · S. Bhardwaj · S. Syam · B. Gonal · P. Saini
CSB-Central Sericultural Research and Training Institute, Pampore 192121, Jammu and Kashmir, India
e-mail: plabaniroyiari@gmail.com

P. Saha
Krishi Vigyan Kendra, West Garo Hills, ICAR Research Complex for North-Eastern Hill Region, Umiam, Meghalaya, India

O. B. Ayoub
College of Temperate Sericulture, Sher-e-Kashmir University of Agricultural Sciences and Technology of Kashmir, Shalimar 190025, Jammu and Kashmir, India

J. A. Parray et al. (eds.), *Sustainable Urban Environment and Waste Management*, Urban Sustainability, https://doi.org/10.1007/978-981-96-1140-9_11

technological innovation, policy support, public education, and community participation. By embracing integrated approaches and prioritizing resource conservation and pollution prevention, cities can move towards more resilient, livable, and sustainable futures.

Keywords Urbanization · By-product utilization · Holistic approaches · Technological innovation · Waste segregation and public awareness

1 Introduction

Urbanization is progressing at an unprecedented rate, leading to significant challenges in solid waste management. As cities expand, so does the volume and complexity of waste generated. Effective solid waste treatment technologies are essential for urban sustainability, encompassing environmental, economic, and social dimensions. This chapter delves into the array of technologies available for solid waste treatment, their implications for urban sustainability, and the challenges faced in implementing these solutions in rapidly growing urban environments. Urban sustainability is a critical issue in modern society, driven by the rapid urbanization and population growth witnessed globally [32]. Effective management of solid waste is one of the most significant challenges within this realm. As urban areas expand, the volume of waste generated increases correspondingly, posing serious environmental, social, and economic consequences. Solid waste treatment technologies are at the forefront of addressing these challenges, aiming to mitigate the adverse impacts of waste on human health and the environment while promoting resource efficiency and sustainability [76]. The surge in solid waste generation in urban areas is attributable to various factors, including industrialization, increased consumerism, and the proliferation of single-use products. Cities, as epicenters of economic activity, produce substantial amounts of waste, ranging from household garbage to industrial by-products. This growth in waste necessitates efficient treatment and management strategies to mitigate pollution, conserve resources, and ensure public health and safety. Traditional methods such as landfilling and incineration have predominated, but they come with limitations and environmental drawbacks, including greenhouse gas emissions, leachate production, and land resource consumption [21].

Innovative solid waste treatment technologies have emerged as pivotal solutions to these challenges. These technologies encompass a range of processes designed to minimize waste, recover valuable resources, and reduce the environmental footprint of waste management activities. Among these, mechanical biological treatment (MBT), anaerobic digestion, and waste-to-energy (WtE) technologies are particularly notable [16].

Mechanical biological treatment (MBT) integrates mechanical sorting and biological processes to separate recyclable materials from organic waste. This method not only reduces the volume of waste destined for landfills but also generates compost and biogas, contributing to resource recovery and energy production [44]. Anaerobic digestion focuses on the biological degradation of organic waste in the absence of oxygen, producing biogas and digestate. This process is especially beneficial for managing food waste and agricultural residues, providing a renewable energy source and organic fertilizer [49].

Waste-to-energy (WtE) technologies, including incineration with energy recovery, gasification, and pyrolysis, convert waste materials into energy through thermal processes. These technologies offer the dual benefits of reducing the volume of waste requiring disposal and generating electricity or heat. However, they also face challenges related to emissions and public acceptance, necessitating stringent regulatory frameworks and advanced pollution control systems [50].

Despite advancements in solid waste treatment technologies, several barriers impede their widespread adoption and effectiveness. Economic constraints, technological limitations, and institutional challenges often hinder the implementation of sustainable waste management practices [26]. Additionally, public awareness and participation are crucial to the success of these technologies. Educating communities about waste segregation, recycling, and the benefits of sustainable waste treatment is essential for fostering an inclusive and effective waste management system [69].

In conclusion, addressing the challenges of solid waste management in urban areas is integral to achieving urban sustainability. The deployment of advanced solid waste treatment technologies presents a viable pathway to mitigate the environmental impacts of waste, recover valuable resources, and promote a circular economy [10]. However, the success of these technologies hinges on overcoming economic, technological, and social barriers, necessitating a collaborative effort from governments, industries, and communities. By embracing innovative waste treatment solutions and fostering a culture of sustainability, urban areas can pave the way towards a more sustainable and resilient future [1].

2 Types of Solid Waste Treatment Technologies

Many of the solid waste management technologies are described as follows (Fig. 1):

2.1 Landfilling

Modern landfills are engineered to minimize environmental impacts through the use of liners, leachate collection systems, and gas management systems. Liners are crucial in preventing contaminants from leaching into the soil and groundwater, while leachate collection systems manage the liquid that percolates through the waste [59]. Gas management systems capture methane, a potent greenhouse gas, which can then be flared or used for energy production. Despite these advancements, landfills pose significant challenges, including extensive land use, potential groundwater

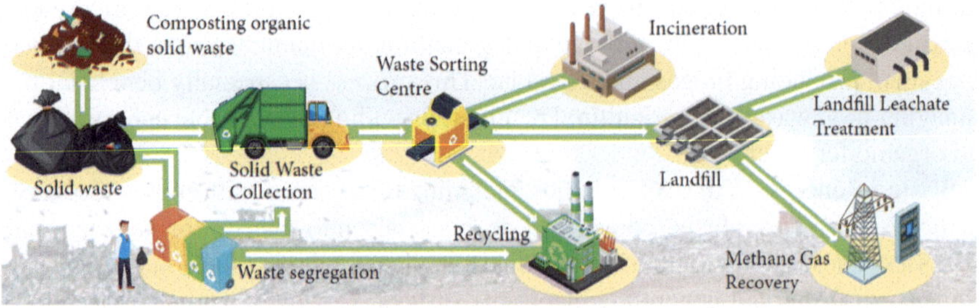

Fig. 1 Management of solid wastes (*Source* Consumers and Industrial Chemistry, Pandai.org)

contamination, and the emission of greenhouse gases [8]. Additionally, they are not a sustainable solution as they merely store waste rather than treating or repurposing it.

Challenges:

- Limited land availability in urban areas.
- Long-term environmental risks.
- Public opposition due to odor and aesthetic concerns.

2.2 *Incineration*

Incineration is a waste treatment technology that involves the combustion of waste materials at high temperatures, typically between 850 and 1,100 °C. This process significantly reduces the volume of waste by up to 90%, converting it into ash, flue gas, and heat. The primary objective of incineration is volume reduction and the stabilization of hazardous components in the waste, although energy recovery has become an increasingly important aspect [34].

Advanced incineration plants, often referred to as Waste-to-Energy (WtE) facilities, are equipped with sophisticated systems to capture and utilize the energy produced during the combustion process [28]. The heat generated is used to produce steam, which can drive turbines to generate electricity or provide district heating. This energy recovery not only offsets the energy consumption of the incineration process itself but also contributes to the local energy grid, enhancing the overall energy efficiency of the waste management system [64].

Modern incineration plants incorporate various pollution control technologies to minimize environmental impacts.

Challenges:

- High capital and operational costs.
- Emission of pollutants, including dioxins and furans.
- Public health concerns and opposition.

2.3 Anaerobic Digestion

Anaerobic digestion (AD) treats organic waste through microbial processes in the absence of oxygen, producing biogas (a mixture of methane and carbon dioxide) and digestate (a nutrient-rich fertilizer). Anaerobic digestion is a biological process that decomposes organic waste in the absence of oxygen [75]. This process produces biogas, a mixture of methane and carbon dioxide, which can be used as a renewable energy source, and digestate, a nutrient-rich substance that can be used as fertilizer [20]. Anaerobic digestion is particularly effective for managing food waste, agricultural residues, and other high-moisture organic wastes. The process reduces the volume of waste, mitigates greenhouse gas emissions by capturing methane, and contributes to a circular economy by transforming waste into valuable resources.

Challenges:

- Requires source-separated organic waste.
- Limited to organic waste streams.
- Technological and operational complexities.

2.4 Waste-to-Energy (WtE)

Waste-to-Energy (WtE) technologies convert waste materials into energy through thermal processes such as incineration, gasification, and pyrolysis [25]. Incineration with energy recovery burns waste at high temperatures, generating electricity or heat while reducing the volume of waste. Gasification and pyrolysis are advanced thermal processes that convert waste into syngas, which can be used to produce electricity, fuels, or chemicals. WtE technologies offer the dual benefits of reducing the volume of waste requiring disposal and generating renewable energy [71]. However, they face challenges related to emissions control, ash management, and public acceptance, necessitating stringent regulatory frameworks and advanced pollution control technologies.

2.5 Composting

Composting is a natural aerobic process that converts organic waste into compost, a valuable soil amendment. Composting is a biological process that decomposes organic waste under aerobic conditions, resulting in the production of compost, a valuable soil amendment [20]. This method is particularly suitable for managing food scraps, yard waste, and other organic materials. Composting can be conducted at various scales, from small household bins to large industrial facilities. The process not only diverts organic waste from landfills but also enriches soil health, reduces the need for chemical fertilizers, and mitigates greenhouse gas emissions by avoiding methane production associated with anaerobic decomposition in landfills [2].

Challenges:

- Requires source separation and collection infrastructure.
- Limited to organic waste.
- Potential for odor and pest issues if not managed properly.

2.6 Mechanical Biological Treatment (MBT)

Mechanical Biological Treatment (MBT) is a comprehensive approach that integrates mechanical sorting and biological processes to manage waste [57]. The mechanical component involves the sorting of recyclable materials and the separation of biodegradable waste from non-recyclable waste. The biological treatment processes, such as composting or anaerobic digestion, then treat the biodegradable fraction. This method reduces the volume of waste sent to landfills and produces valuable by-products such as compost and biogas [46]. MBT facilities can be designed to process mixed municipal solid waste or source-separated organic waste, enhancing the efficiency of recycling and resource recovery efforts [61, 62].

Challenges:

- High initial investment and operational costs.
- Complex technology requiring skilled labor.
- Varies in efficiency depending on waste composition.

2.7 Recycling

Recycling involves converting waste materials into new products, conserving resources, and reducing the need for raw materials. Recycling involves the collection, sorting, and processing of materials that can be reintroduced into the manufacturing process [9]. Commonly recycled materials include paper, cardboard, plastics, glass, and metals. Recycling reduces the demand for raw materials, conserves energy, and minimizes the environmental impact of resource extraction and processing [41]. Effective recycling programs require robust infrastructure, public participation, and market demand for recycled materials. Challenges in recycling include contamination of recyclables, fluctuating market prices, and the need for efficient sorting and processing technologies [11].

Challenges:

- Requires effective sorting and separation.
- Contamination of recyclables.
- Market fluctuations for recyclable materials.

The diverse array of solid waste treatment technologies each plays a vital role in managing urban waste sustainably [30]. While traditional methods like landfilling continue to be prevalent, innovative approaches such as MBT, anaerobic digestion, WtE, composting, and recycling offer promising alternatives that contribute to resource recovery, renewable energy production, and environmental protection [13]. The effective integration and implementation of these technologies are essential for advancing urban sustainability and addressing the growing challenges of waste management [17].

3 Urban Sustainability and Solid Waste Treatment

Urban sustainability is intrinsically linked to effective solid waste management. By integrating environmental, economic, and social dimensions, cities can develop comprehensive waste treatment strategies that mitigate environmental impacts, promote economic growth, and enhance social well-being [15, 37]. The implementation of advanced technologies, coupled with robust public participation and education, is essential for fostering sustainable urban environments [70].

3.1 Environmental Sustainability

Effective solid waste management is crucial for environmental sustainability. By employing advanced technologies and strategies, urban areas can significantly reduce their environmental impacts [43]. Key approaches include:

Pollution Minimization: Proper waste management prevents litter, reduces air and water pollution, and mitigates the spread of hazardous substances [12]. Technologies like advanced landfill liners and leachate treatment systems help safeguard soil and groundwater from contamination [27, 61, 62].

Resource Conservation: Recycling and composting divert substantial amounts of waste from landfills, preserving natural resources and reducing the need for raw material extraction. Recycling processes recover valuable materials such as metals, plastics, and paper, which can be reintroduced into the manufacturing cycle, thereby conserving resources and reducing energy consumption [35].

Greenhouse Gas Emissions Reduction: Solid waste management practices such as composting and anaerobic digestion reduce methane emissions from organic waste [48]. Additionally, waste-to-energy (WtE) facilities can offset fossil fuel use by generating renewable energy from waste, thereby contributing to lower carbon emissions [4, 31].

3.2 Economic Sustainability

Solid waste treatment technologies can contribute to economic sustainability by creating jobs, generating energy, and recovering valuable materials. However, the high costs associated with advanced technologies like incineration and MBT necessitate careful economic planning and investment [71].

Social Sustainability

Urban waste management systems must address social dimensions to achieve comprehensive sustainability. Key considerations include:

- **Equitable Access to Waste Services**: Ensuring that all urban residents, regardless of socioeconomic status, have access to effective waste collection and treatment services is vital. This reduces disparities in waste-related health and environmental impacts [6].
- **Health Impact Minimization**: Proper waste management reduces the health risks associated with improper waste disposal, such as exposure to hazardous substances and the proliferation of disease vectors [14]. Advanced technologies and efficient waste collection systems play a crucial role in protecting public health.
- **Public Participation and Education**: The success of recycling and composting programs heavily relies on active public participation [22]. Educating communities about the importance of waste segregation, recycling, and composting fosters a culture of sustainability and increases the effectiveness of waste management programs [63]. Public awareness campaigns and community engagement initiatives are essential for encouraging responsible waste practices [55].

3.3 Economic Sustainability

Solid waste treatment technologies can significantly contribute to economic sustainability by:

- **Job Creation**: The development and operation of waste management facilities, including recycling plants, composting sites, and WtE plants, create numerous employment opportunities in collection, processing, and maintenance roles [3].
- **Energy Generation**: WtE technologies harness the energy potential of waste, producing electricity and heat. This not only reduces the demand for non-renewable energy sources but also can provide a revenue stream through the sale of generated energy [29].
- **Material Recovery**: Recycling and MBT processes recover valuable materials from waste, which can be sold and reused, thereby generating economic value [36]. However, these technologies require significant initial investments and ongoing operational costs. Careful economic planning and investment strategies are essential to ensure that the benefits outweigh the costs [68].

4 Challenges in Implementing Solid Waste Treatment Technologies

4.1 Policy and Regulatory Challenges

Effective solid waste management necessitates robust regulatory frameworks and supportive policies that promote sustainable practices [47, 74]. However, many cities face significant policy and regulatory challenges:

- **Inadequate Regulations**: In numerous urban areas, the regulations governing waste management are outdated, insufficient, or poorly designed [67]. This inadequacy makes it difficult to enforce standards that are necessary for the adoption of advanced waste treatment technologies.
- **Lack of Enforcement**: Even when regulations exist, the lack of enforcement mechanisms can undermine their effectiveness. Without stringent oversight, illegal dumping and improper waste handling practices persist, negating efforts to implement sustainable waste treatment technologies [45].
- **Inconsistent Policies**: Policies related to waste management often vary significantly between jurisdictions, leading to inconsistencies that complicate the adoption of new technologies [39]. Harmonizing these policies and ensuring they align with best practices is crucial for effective waste management.

4.2 Financial and Economic Challenges

The financial and economic barriers to implementing advanced waste treatment technologies are considerable:

- **High Initial Costs**: The establishment of facilities for advanced technologies like mechanical biological treatment (MBT), waste-to-energy (WtE), and anaerobic digestion involves substantial capital investment [5]. These high initial costs can be prohibitive for many municipalities, especially in developing regions.
- **Operational and Maintenance Costs**: Beyond the initial setup, the ongoing operational and maintenance expenses of advanced waste treatment facilities are significant. Ensuring consistent funding for these activities is a major challenge.
- **Innovative Financing Mechanisms**: Addressing financial challenges requires innovative solutions such as public–private partnerships (PPPs), where private investment is leveraged to support public waste management initiatives [54, 72]. Additionally, mechanisms like green bonds and international funding programs can provide necessary capital.

4.3 Technical and Operational Challenges

Urban areas encounter several technical and operational obstacles in managing waste effectively:

- **Collection Logistics**: Efficient waste collection is fundamental to effective waste management [60]. Urban areas, particularly those with dense populations and limited infrastructure, face logistical challenges in waste collection, including transportation bottlenecks and inadequate collection coverage [73].
- **Technological Integration**: Integrating advanced technologies with existing waste management systems can be complex. It requires compatibility between new and old systems, as well as the ability to handle diverse types of waste effectively [7, 38].
- **Skilled Labor**: Operating and maintaining advanced waste treatment facilities require a skilled workforce [66]. However, there is often a shortage of trained personnel, necessitating ongoing capacity-building efforts to develop the required skills and expertise.
- **Continuous Innovation**: Keeping pace with technological advancements in waste management demands continuous innovation [42]. Urban areas need to invest in research and development to adopt the latest technologies and improve existing systems [53].

4.4 Social and Behavioral Challenges

Public awareness and participation are critical for the success of waste management programs, yet social and behavioral challenges often impede progress:

- **Awareness and Education**: Many residents lack awareness of the importance of proper waste management practices [18]. Comprehensive education campaigns are essential to inform the public about waste reduction, separation, and recycling.
- **Behavioral Change**: Encouraging behavioral change is challenging. People may be resistant to altering their habits, such as segregating waste at the source or participating in recycling programs [24]. Sustained efforts are required to change public attitudes and behaviors towards waste management.
- **Community Engagement**: Engaging communities in waste management programs ensures higher participation rates and compliance [40]. This involves not only educating the public but also actively involving them in decision-making processes and program implementation.
- **Equity and Accessibility**: Ensuring that all community members have equitable access to waste management services is vital [51]. This includes addressing disparities in service provision to marginalized or underserved populations and ensuring that programs are inclusive and accessible to all [30].

The implementation of solid waste treatment technologies in urban areas is fraught with challenges spanning policy, financial, technical, and social dimensions [52]. Overcoming these challenges requires a multifaceted approach that includes robust regulatory frameworks, innovative financing mechanisms, continuous technical innovation, and comprehensive public education and engagement [19]. By addressing these barriers, cities can enhance their waste management systems and move towards more sustainable urban environments [27].

CASE STUDIES

1. Singapore's Integrated Waste Management System

Singapore has gained global recognition for its highly efficient and sustainable approach to solid waste management, integrating incineration, recycling, and landfilling into a cohesive system that maximizes resource recovery and minimizes environmental impact (Fig. 2). At the forefront of this system is the Tuas South Incineration Plant (TSIP), Singapore's largest waste-to-energy facility. TSIP processes over 3,000 tons of waste daily, converting it into energy that contributes significantly to the national grid. Advanced flue gas treatment systems at TSIP capture pollutants such as dioxins, furans, and heavy metals, ensuring that emissions meet stringent environmental standards. Moreover, the plant incorporates technologies for recovering metals from incineration bottom ash through magnetic and eddy current separators, thereby reducing the demand for virgin raw materials and promoting sustainable resource management [65].

Complementing the energy recovery from incineration, Singapore has implemented comprehensive recycling programs aimed at maximizing the reuse of materials. Public awareness campaigns and widespread provision of recycling bins encourage citizens to segregate waste effectively. Material Recovery Facilities (MRFs) play a crucial role in sorting and preparing recyclables for processing into new materials, further closing the loop on resource consumption. Regulatory measures, such as mandatory recycling for certain sectors and Extended Producer Responsibility (EPR) schemes, incentivize businesses to manage their products' end-of-life responsibly, thereby reducing overall waste generation and promoting a circular economy.

In tandem with these efforts, Singapore manages residual waste through the Semakau Landfill, a pioneering offshore landfill. Constructed with environmental sustainability in mind, Semakau Landfill utilizes impermeable membranes and advanced leachate collection systems to prevent contamination of surrounding marine ecosystems. A portion of the landfill has been dedicated to conservation and recreational activities, showcasing how landfill sites can coexist harmoniously with nature while serving essential waste management functions [56].

The integrated approach to waste management in Singapore not only enhances resource efficiency and environmental protection but also supports broader sustainability goals. By reducing reliance on landfills and maximizing energy and material recovery, Singapore minimizes its environmental footprint and contributes to

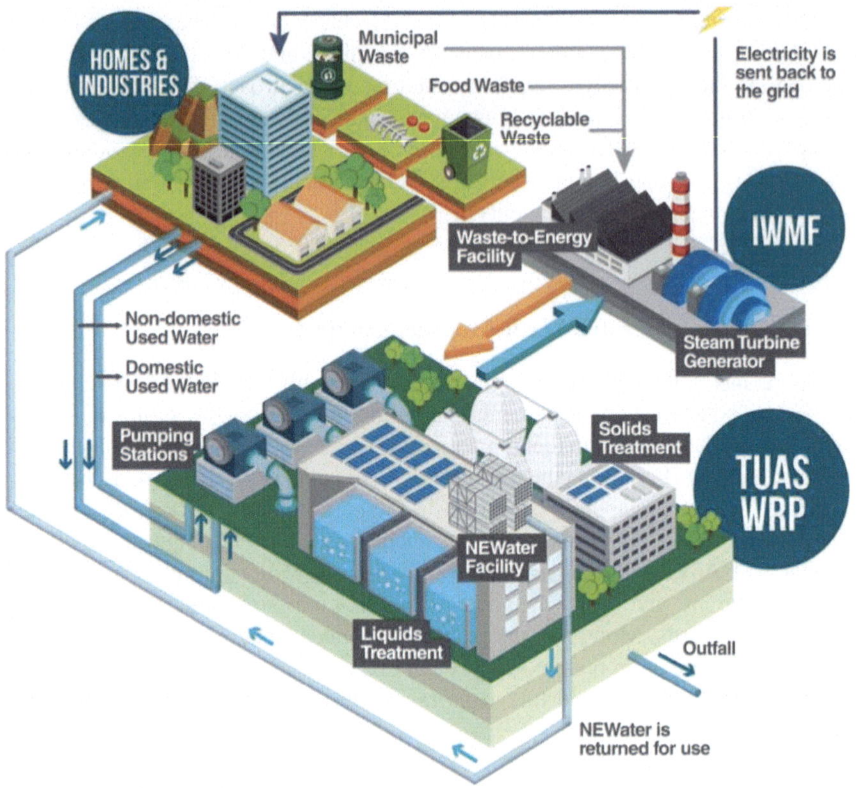

Fig. 2 Integrated waste management facility at Tuas view Basin, Singapore (*Source* Black and Veatch Company)

global efforts in combating climate change. Ongoing advancements in technology and continual improvement in waste management practices underscore Singapore's commitment to innovation and sustainability, setting a benchmark for cities worldwide seeking effective solutions to the challenges of urban waste management in the twenty-first century.

2. **San Francisco's Zero Waste Program**

San Francisco's Zero Waste Program stands as a pioneering initiative aimed at achieving the ambitious goal of zero waste by 2020. This program represents a comprehensive approach integrating aggressive recycling, composting initiatives, stringent policies, and extensive public education campaigns. At its core, the program emphasizes reducing waste sent to landfills by maximizing material recovery and promoting sustainable consumption practices [58].

Central to San Francisco's strategy is its robust recycling program, which encourages residents and businesses to segregate recyclables effectively. The city provides easily accessible recycling bins and educates the community on the importance of recycling to conserve resources and reduce environmental impact. Material Recovery Facilities (MRFs) play a critical role in this process by sorting and processing collected recyclables, ensuring they are efficiently reintegrated into the manufacturing supply chain [58].

Complementing recycling efforts, San Francisco has implemented one of the most extensive composting programs in the United States. Food scraps, yard trimmings, and other organic waste materials are collected separately and sent to composting facilities. This not only diverts significant amounts of organic waste from landfills but also produces high-quality compost used to enrich soil and support local agriculture and landscaping initiatives.

The success of San Francisco's Zero Waste Program is also attributable to its stringent policies and regulations. The city has implemented mandatory recycling and composting ordinances, requiring residents and businesses to comply with waste separation guidelines. These policies are reinforced through fines and penalties for non-compliance, ensuring widespread adherence to waste reduction practices.

Furthermore, public education and community engagement are pivotal components of the program's success. San Francisco conducts extensive outreach campaigns, workshops, and educational programs to raise awareness about waste reduction strategies and foster a culture of environmental stewardship. Community partnerships with local businesses, schools, and nonprofit organizations play a crucial role in promoting sustainable practices and encouraging participation in recycling and composting initiatives.

As San Francisco approaches its target year of 2020, the city's Zero Waste Program serves as a model of effective urban waste management. By prioritizing resource recovery, promoting sustainable behaviors, and leveraging policy frameworks and community engagement, San Francisco demonstrates that achieving zero waste is not only aspirational but also achievable through strategic planning, innovation, and collective action. The program's outcomes underscore the importance of local leadership, innovation, and collaboration in addressing global environmental challenges and advancing towards a more sustainable future for urban communities worldwide.

3. **Germany's Recycling System**

Germany's recycling system stands as a global 8exemplar of efficiency and environmental stewardship, underpinned by the Green Dot program and a comprehensive regulatory framework (Fig. 3). Central to its success is the Green Dot symbol, which signifies a commitment by manufacturers to finance the recovery and recycling of packaging materials. Established in 1990, this program mandates that producers take extended producer responsibility (EPR) for their packaging, ensuring that they fund and participate in the recycling infrastructure that manages their products' end-of-life stages [33].

Germany's recycling success is further bolstered by strict regulations that mandate recycling targets and waste management practices. These regulations incentivize eco-design, pushing manufacturers to use recyclable materials and minimize packaging

Fig. 3 Comprehensive guide to trash sorting in Germany (*Source* https://ftzohu.medium.com/recycling-in-germany-guide-to-trash-sorting-dcbcd10f0fe6)

waste. Companies must comply with stringent requirements or face financial penalties, which drives widespread participation in recycling initiatives and ensures robust funding for recycling infrastructure nationwide.

The country boasts a highly efficient waste collection infrastructure that facilitates the sorting and collection of various waste streams. Residents and businesses are rigorously encouraged to separate their waste into categories like paper, plastics, metals, and biowaste, making it easier to recycle and recover valuable resources. Local municipalities coordinate collection services, ensuring that materials are efficiently transported to recycling facilities where they undergo sorting and processing [23].

Germany's commitment to recycling has yielded impressive results, with over 67% of packaging waste recycled annually. This achievement significantly reduces the amount of waste sent to landfills or incineration, conserving natural resources and mitigating environmental impacts such as greenhouse gas emissions and pollution.

Looking ahead, Germany continues to innovate in waste management and recycling technologies. Research and development efforts focus on improving recycling processes, developing new technologies for challenging materials, and advancing the circular economy principles. Initiatives like bottle deposit systems and efforts to reduce single-use plastics further underscore Germany's proactive approach to sustainable consumption and waste reduction.

Germany's recycling system demonstrates the effectiveness of integrating strict regulations, extended producer responsibility, and a robust infrastructure to achieve high recycling rates and environmental sustainability. By prioritizing resource conservation, reducing landfill waste, and promoting circular economy principles, Germany serves as a global model for responsible waste management, illustrating how ambitious recycling goals can be realized through comprehensive planning, regulatory frameworks, and public engagement.

5 Conclusion

Achieving urban sustainability through effective solid waste management demands a holistic and adaptive approach that integrates advanced technologies, supportive policies, economic viability, technical feasibility, and robust public engagement. Cities worldwide face diverse challenges in managing their waste streams, exacerbated by rapid urbanization, population growth, and evolving consumption patterns. To address these challenges, cities must adopt a strategic framework that balances environmental stewardship with economic efficiency and social equity. Advanced technologies such as waste-to-energy incineration, anaerobic digestion for organic waste, and sophisticated recycling processes offer promising solutions to reduce waste volumes, recover resources, and minimize environmental impacts. These technologies should be implemented alongside supportive policies that set clear targets for waste reduction, recycling rates, and the promotion of circular economy principles. Policies promoting extended producer responsibility (EPR), landfill diversion incentives, and sustainable procurement practices can incentivize businesses and industries to adopt environmentally friendly practices throughout their product life cycles. Furthermore, economic viability is essential in ensuring the sustainability of waste management solutions. Investments in infrastructure and technology must be justified by long-term cost savings, environmental benefits, and societal well-being. Cities can explore public–private partnerships, innovative financing mechanisms, and incentives for green investments to support the transition towards sustainable waste management systems.

Technical feasibility plays a crucial role in the practical implementation of waste management solutions within urban settings. Factors such as space constraints, technological compatibility, and environmental regulations must be carefully considered to optimize the effectiveness of waste treatment facilities and recycling programs. Research and development in waste management technologies should continue to drive innovation and improve efficiency in resource recovery and waste processing.

Moreover, public participation and awareness are fundamental to the success of sustainable waste management initiatives. Educating residents and businesses about waste reduction strategies, recycling practices, and the importance of environmental conservation can foster a culture of responsibility and collective action. Community engagement initiatives, including outreach campaigns, educational programs, and participatory decision-making processes, empower citizens to contribute actively to waste reduction efforts and environmental sustainability goals.

In essence, the journey towards urban sustainability through solid waste management requires a collaborative effort across multiple sectors and stakeholders. By embracing innovative technologies, implementing supportive policies, ensuring economic viability, assessing technical feasibility, and engaging the public, cities can effectively manage their waste streams, minimize environmental footprints, and enhance resilience against future challenges. This comprehensive approach not only safeguards natural resources and ecosystems but also promotes equitable and prosperous urban environments for current and future generations. By committing to these principles, cities can pave the way towards a sustainable and resilient future.

References

1. Abdel-Shafy HI, Mansour MS (2018) Solid waste issue: sources, composition, disposal, recycling, and valorization. Egypt J Pet 27(4):1275–1290
2. Abubakar IR, Maniruzzaman KM, Dano UL, AlShihri FS, AlShammari MS, Ahmed SM, Al-Gehlani WA, Alrawaf TI (2022) Environmental sustainability impacts of solid waste management practices in the global South. Int J Environ Res Public Health 19(19):12717
3. Aleluia J, Ferrão P (2017) Assessing the costs of municipal solid waste treatment technologies in developing Asian countries. Waste Manag 1(69):592–608
4. Ambaye TG, Djellabi R, Vaccari M, Prasad S, Aminabhavi TM, Rtimi S (2023) Emerging technologies and sustainable strategies for municipal solid waste valorization: challenges of circular economy implementation. J Clean Prod 15(423):138708
5. Amin N, Aslam M, Yasin M, Hossain S, Shahid MK, Inayat A, Samir A, Ahmad R, Murshed MN, Khurram MS, El Sayed ME (2023) Municipal solid waste treatment for bioenergy and resource production: potential technologies, techno-economic-environmental aspects and implications of membrane-based recovery. Chemosphere 1(323):138196
6. Arteaga C, Silva J, Yarasca-Aybar C (2023) Solid waste management and urban environmental quality of public space in Chiclayo, Peru. City Environ Interact 1(20):100112
7. Azevedo BD, Scavarda LF, Caiado RG (2019) Urban solid waste management in developing countries from the sustainable supply chain management perspective: a case study of Brazil's largest slum. J Clean Prod 1(233):1377–1386
8. Batista M, Caiado RG, Quelhas OL, Lima GB, Leal Filho W, Yparraguirre IT (2021) A framework for sustainable and integrated municipal solid waste management: barriers and critical factors to developing countries. J Clean Prod 20(312):127516
9. Bharathiraja B, Sudharsana T, Jayamuthunagai J, Praveenkumar R, Chozhavendhan S, Iyyappan J (2018) Biogas production—a review on composition, fuel properties, feed stock and principles of anaerobic digestion. Renew Sustain Energy Rev 90(April):570–582
10. Bui TD, Tseng JW, Tseng ML, Lim MK (2022) Opportunities and challenges for solid waste reuse and recycling in emerging economies: a hybrid analysis. Resour Conserv Recycl 1(177):105968
11. Cao Y, Fu L, Mofrad A (2019) Combined-gasification of biomass and municipal solid waste in a fluidized bed gasifier. J Energy Inst 92(6):1683–1688
12. Capodaglio AG, Callegari A, Cecconet D, Molognoni D (2017) Sustainability of decentralized wastewater treatment technologies. Water Pract Technol 12(2):463–477
13. Cheng H, Hu Y (2010) Municipal solid waste (MSW) as a renewable source of energy: current and future practices in China. Biores Technol 101(11):3816–3824
14. Chien CF, Aviso K, Tseng ML, Fujii M, Lim MK (2023) Solid waste management in emerging economies: opportunities and challenges for reuse and recycling. Resour Conserv Recycl 1(188):106635
15. Cohen B (2004) Urban growth in developing countries: a review of current trends and a caution regarding existing forecasts. World Dev 32(1):23–51
16. da Silva L, Prietto PD, Korf EP (2019) Sustainability indicators for urban solid waste management in large and medium-sized worldwide cities. J Clean Prod 10(237):117802

17. Dada MA, Obaigbena A, Majemite MT, Oliha JS, Biu PW (2024) Innovative approaches to waste resource management: implications for environmental sustainability and policy. Eng Sci Technol J 5(1):115–127
18. Dadario N, Gabriel Filho LR, Cremasco CP, Santos FA, Rizk MC, Mollo NM (2023) Waste-to-energy recovery from municipal solid waste: global scenario and prospects of mass burning technology in Brazil. Sustainability 15(6):5397
19. Das S, Lee SH, Kumar P, Kim KH, Lee SS, Bhattacharya SS (2019) Solid waste management: scope and the challenge of sustainability. J Clean Prod 10(228):658–678
20. David VE Jr, John Y, Hussain S (2020) Rethinking sustainability: a review of Liberia's municipal solid waste management systems, status, and challenges. J Mater Cycles Waste Manag 22(5):1299–1317
21. de Oliveira Leite FF, Palacio JC, Batista MJ, Renó ML (2022) Evaluation of technological alternatives for the treatment of urban solid waste: a case study of Minas Gerais, Brazil. J Clean Prod 1(330):129618
22. de Souza Melaré AV, González SM, Faceli K, Casadei V (2017) Technologies and decision support systems to aid solid-waste management: a systematic review. Waste Manag 1(59):567–584
23. Dehio J, Janßen-Timmen R, Rothgang M (2023) Regulating markets for post-consumer recycling plastics: experiences from Germany's dual system. Resour Conserv Recycl 1(196):107048
24. Ebekozien A, Samsurijan MS, Aigbavboa C, Malek NM (2023) Malaysia's low-cost housing solid waste management in the era of COVID-19: the role of fourth industrial revolution technologies. Manag Environ Qual An Int J 34(1):80–98
25. Ferdoush MR, Al Aziz R, Karmaker CL, Debnath B, Limon MH, Bari AM (2024) Unraveling the challenges of waste-to-energy transition in emerging economies: implications for sustainability. Innov Green Dev 3(2):100121
26. Ferronato N, Gorritty Portillo MA, Guisbert Lizarazu EG, Torretta V, Bezzi M, Ragazzi M (2018) The municipal solid waste management of La Paz (Bolivia): challenges and opportunities for a sustainable development. Waste Manag Res 36(3):288–299
27. Fuss M, Barros RT, Poganietz WR (2018) Designing a framework for municipal solid waste management towards sustainability in emerging economy countries—an application to a case study in Belo Horizonte (Brazil). J Clean Prod 20(178):655–664
28. Heidari R, Yazdanparast R, Jabbarzadeh A (2019) Sustainable design of a municipal solid waste management system considering waste separators: a real-world application. Sustain Cities Soc 1(47):101457
29. Hettiarachchi H, Ryu S, Caucci S, Silva R (2018) Municipal solid waste management in Latin America and the Caribbean: Issues and potential solutions from the governance perspective. Recycling 3(2):19
30. Hossain I, Haque AM, Ullah SA (2024) Assessing sustainable waste management practices in Rajshahi City corporation: an analysis for local government enhancement using IoT, AI, and Android technology. Environ Sci Pollut Res 6:1–9
31. Ike CC, Ezeibe CC, Anijiofor SC, Daud NN (2018) Solid waste management in Nigeria: problems, prospects, and policies. J Solid Waste Technol Manag 44(2):163–172
32. Khan S, Anjum R, Raza ST, Bazai NA, Ihtisham M (2022) Technologies for municipal solid waste management: current status, challenges, and future perspectives. Chemosphere 1(288):132403
33. Kullmann F, Markewitz P, Kotzur L, Stolten D (2022) The value of recycling for low-carbon energy systems—a case study of Germany's energy transition. Energy 1(256):124660
34. Kumar A, Agrawal A (2020) Recent trends in solid waste management status, challenges, and potential for the future Indian cities—a review. Curr Res Environ Sustain 1(2):100011
35. Kumar A, Singh E, Mishra R, Lo SL, Kumar S (2023) Global trends in municipal solid waste treatment technologies through the lens of sustainable energy development opportunity. Energy 15(275):127471
36. Kumari T, Raghubanshi AS (2023) Waste management practices in the developing nations: challenges and opportunities. Waste Manag Resour Recycl Dev World 1:773–797

37. Kwakye SO, Amuah EE, Ankoma KA, Agyemang EB, Owusu BG (2024) Understanding the performance and challenges of solid waste management in an emerging megacity: insights from the developing world. Environ Chall 1(14):100805
38. Lakiotia EN, Moustakas KG, Komilisc DP, Domopouloua AE, Karayannisa VG (2017) Sustainable solid waste management: socio-economic considerations. Chem Eng 56
39. Li CH, Lee TT, Lau SS (2023) Enhancement of municipal solid waste Management in Hong Kong through innovative solutions: a review. Sustainability 15(4):3310
40. Lino FA, Ismail KA, Castañeda-Ayarza JA (2023) Municipal solid waste treatment in Brazil: a comprehensive review. Energy Nexus 9:100232
41. Lombardi L, Castaldi MJ (2024) Energy recovery from residual municipal solid waste: state of the art and perspectives within the challenge to climate change. Energies 17(2):395
42. Marello M, Helwege A (2018) Solid waste management and social inclusion of wastepickers: opportunities and challenges. Lat Am Perspect 45(1):108–129
43. Meena MD, Dotaniya ML, Meena BL, Rai PK, Antil RS, Meena HS, Meena LK, Dotaniya CK, Meena VS, Ghosh A, Meena KN (2023) Municipal solid waste: opportunities, challenges and management policies in India: a review. Waste Manag Bull 1(1):4–18
44. Mor S, Ravindra K (2023) Municipal solid waste landfills in lower-and middle-income countries: environmental impacts, challenges and sustainable management practices. Process Saf Environ Prot 1(174):510–530
45. Munir MT, Li B, Naqvi M (2023) Revolutionizing municipal solid waste management (MSWM) with machine learning as a clean resource: Opportunities, challenges and solutions. Fuel 15(348):128548
46. Nanda S, Berruti F (2021) Municipal solid waste management and landfilling technologies: a review. Environ Chem Lett 19(2):1433–1456
47. Narayanamoorthy S, Anuja A, Pragathi S, Sandra M, Ferrara M, Ahmadian A, Kang D (2024) Assessment of inorganic solid waste management techniques using full consistency and extended MABAC method. Environ Sci Pollut Res 31(7):9981–9991
48. Oliveira Santos MA, Leite Soares FI, Bizerra da Costa M, Alves de Souza Paes DC, de Almeida Nascimento MA (2024) Challenges and perspectives in the management of solid urban waste in municipalities in para state: a critical and multifaceted analysis. Environ Soc Manag J Revista de Gestão Social e Ambiental 18(3)
49. Pires A, Martinho G, Rodrigues S, Gomes MI (2019) Sustainable solid waste collection and management. Springer International Publishing, Cham, pp 45–59
50. Ravichandran C, Venkatesan G (2021) Toward sustainable solid waste management–challenges and opportunities. In: Handbook of advanced approaches towards pollution prevention and control, pp 67–103
51. Ren J, Liang H, Chan FT (2017) Urban sewage sludge, sustainability, and transition for eco-city: multi-criteria sustainability assessment of technologies based on best-worst method. Technol Forecast Soc Chang 1(116):29–39
52. Rodić L, Wilson DC (2017) Resolving governance issues to achieve priority sustainable development goals related to solid waste management in developing countries. Sustainability 9(3):404
53. Saadatlu EA, Barzinpour F, Yaghoubi S (2023) A sustainable municipal solid waste system under leachate treatment impact along with leakage control and source separation. Process Saf Environ Prot 1(169):982–998
54. Salem KS, Clayson K, Salas M, Haque N, Rao R, Agate S, Singh A, Levis JW, Mittal A, Yarbrough JM, Venditti R (2023) A critical review of existing and emerging technologies and systems to optimize solid waste management for feedstocks and energy conversion. Matter
55. Sharma KD, Jain S (2019) Overview of municipal solid waste generation, composition, and management in India. J Environ Eng 145(3):04018143
56. Shekdar AV (2009) Sustainable solid waste management: an integrated approach for Asian countries. Waste Manag 29(4):1438–1448
57. Shovon SM, Akash FA, Rahman W, Rahman MA, Chakraborty P, Hossain HZ, Monir MU (2024) Strategies of managing solid waste and energy recovery for a developing country—a review. Heliyon

58. Silva A, Rosano M, Stocker L, Gorissen L (2017) From waste to sustainable materials management: three case studies of the transition journey. Waste Manag 1(61):547–557
59. Simatele DM, Dlamini S, Kubanza NS (2017) From informality to formality: perspectives on the challenges of integrating solid waste management into the urban development and planning policy in Johannesburg, South Africa. Habitat Int 1(63):122–130
60. Singh A (2019) Managing the uncertainty problems of municipal solid waste disposal. J Environ Manag 15(240):259–265
61. Singh D, Dikshit AK, Kumar S (2024) Smart technological options in collection and transportation of municipal solid waste in urban areas: a mini review. Waste Manag Res 42(1):3–15
62. Singh M, Singh M, Singh SK (2024) Tackling municipal solid waste crisis in India: insights into cutting-edge technologies and risk assessment. Sci Total Environ 29:170453
63. Soni A, Das PK, Kumar P (2023) A review on the municipal solid waste management status, challenges and potential for the future Indian cities. Environ Dev Sustain 25(12):13755–13803
64. Soni A, Das PK, Hashmi AW, Yusuf M, Kamyab H, Chelliapan S (2022) Challenges and opportunities of utilizing municipal solid waste as alternative building materials for sustainable development goals: a review. Sustain Chem Pharm 1(27):100706
65. Syme R (2021) A review of domestic waste management policy and law in Singapore. Asia Pacific J Environ Law 24(1):90–119
66. Szpilko D, de la Torre GA, Jimenez Naharro F, Rzepka A, Remiszewska A (2023) Waste management in the smart city: current practices and future directions. Resources 12(10):115
67. Tarpani RR, Azapagic A (2023) Life cycle sustainability assessment of advanced treatment techniques for urban wastewater reuse and sewage sludge resource recovery. Sci Total Environ 15(869):161771
68. Trulli E, Ferronato N, Torretta V, Piscitelli M, Masi S, Mancini I (2018) Sustainable mechanical biological treatment of solid waste in urbanized areas with low recycling rates. Waste Manag 1(71):556–564
69. Tsui TH, Wong JW (2019) A critical review: emerging bioeconomy and waste-to-energy technologies for sustainable municipal solid waste management. Waste Disposal Sustain Energy 1:151–167
70. Verma P, Raghubanshi AS (2018) Urban sustainability indicators: challenges and opportunities. Ecol Ind 1(93):282–291
71. Voukkali I, Papamichael I, Loizia P, Zorpas AA (2024) Urbanization and solid waste production: prospects and challenges. Environ Sci Pollut Res 31(12):17678–17689
72. Wei Y, Li J, Shi D, Liu G, Zhao Y, Shimaoka T (2017) Environmental challenges impeding the composting of biodegradable municipal solid waste: a critical review. Resour Conserv Recycl 1(122):51–65
73. Xiao W, Liu T, Tong X (2023) Assessing the carbon reduction potential of municipal solid waste management transition: Effects of incineration, technology and sorting in Chinese cities. Resour Conserv Recycl 1(188):106713
74. Ye J, Chen X, Chen C, Bate B (2019) Emerging sustainable technologies for remediation of soils and groundwater in a municipal solid waste landfill site—a review. Chemosphere 1(227):681–702
75. Yong ZJ, Bashir MJ, Ng CA, Sethupathi S, Lim JW, Show PL (2019) Sustainable waste-to-energy development in Malaysia: appraisal of environmental, financial, and public issues related with energy recovery from municipal solid waste. Processes 7(10):676
76. Zohoori M, Ghani A (2017) Municipal solid waste management challenges and problems for cities in low-income and developing countries. Int J Sci Eng Appl 6(2):39–48

Mulberry Trees: A Sustainable Solution for Urban Forestry and Improved Air Quality

Gulab Khan Rohela, Pawan Saini, S. Syam, Plabani Roy, Basanagouda Gonal, Danishta Aziz, Omais Bin Ayoub, and M. Maheswari

Abstract Mulberry trees (*Morus* spp.) are emerging as ideal candidates for urban forestry, playing a significant role in enhancing air quality and fostering sustainable urban development. This abstract explores the multifaceted benefits of mulberries in urban settings, highlighting their adaptability, resilience, and ecological contributions. Mulberry trees are known for their fast growth and ability to thrive in various urban environments, making them suitable for city landscapes. Their dense foliage effectively captures airborne particulate matter and absorbs carbon dioxide, thereby contributing to cleaner air. The robust root systems of mulberries prevent soil erosion and promote soil health, further supporting urban sustainability. Beyond environmental benefits, mulberry trees offer social and aesthetic advantages. They provide shade, reducing the urban heat island effect, and contribute to the visual appeal of urban areas with their lush greenery. Additionally, mulberries produce edible fruit, serving as a food source for both urban wildlife and city residents interested in urban agriculture. Given their ecological, social, and aesthetic benefits, mulberry trees are an ideal choice for urban forestry initiatives aimed at improving air quality and supporting sustainable development in cities. This book chapter underscores the importance of incorporating mulberries into urban planning to create greener, healthier, and more sustainable urban environments.

Keywords Adaptability · City landscapes · Resilience · Ecological contributions · Urban sustainability

G. K. Rohela (✉) · P. Saini · S. Syam · P. Roy · B. Gonal · M. Maheswari
CSB-Central Sericultural Research & Training Institute (CSR&TI), Pampore 192 121, Jammu-Kashmir (UT), India
e-mail: gulab_biotech@yahoo.co.in

D. Aziz
Division of Entomology, Faculty of Agriculture, Sher-e-Kashmir University of Agricultural Science and Technology – Kashmir (SKUAST-K), Sopore, Wadura, Baramulla, Jammu and Kashmir, India

O. B. Ayoub
College of Temperate Sericulture, Sher-e-Kashmir University of Agricultural Science and Technology of Kashmir (SKUAST-K), Shalimar 190 025, Jammu and Kashmir, India

J. A. Parray et al. (eds.), *Sustainable Urban Environment and Waste Management*, Urban Sustainability, https://doi.org/10.1007/978-981-96-1140-9_12

1 Introduction

Urbanization, while often celebrated as a hallmark of economic progress and societal development, also brings with it a suite of environmental challenges that pose significant threats to the quality of life in cities [7]. The unprecedented rate of urban growth has been accompanied by a range of issues, including air pollution, loss of green spaces, and the consequent decline in urban biodiversity [14]. These challenges not only affect environmental sustainability but also have serious implications for public health and the well-being of urban residents [18]. One of the most pressing environmental challenges associated with urbanization is air pollution. As cities expand, the increase in vehicular traffic, industrial activities, and construction leads to the release of harmful pollutants such as particulate matter (PM), nitrogen oxides (NOx), sulfur dioxide (SO_2), and volatile organic compounds (VOCs) into the atmosphere [43]. According to a study by Ghorani-Azam et al. [8], air pollution is a major environmental risk to health, contributing to respiratory diseases, cardiovascular problems, and premature death. The World Health Organization (WHO) estimates that air pollution is responsible for approximately 7 million premature deaths annually, with urban areas bearing the brunt of this burden [48].

Another significant issue is the loss of green spaces. Urbanization typically involves the conversion of natural landscapes into built environments, reducing the availability of parks, gardens, and forests that provide essential ecological services. These green spaces act as the "lungs" of cities, helping to filter air, regulate temperatures, and maintain ecological balance. The loss of these areas not only degrades air quality but also diminishes the psychological and physical health benefits associated with access to nature [47]. Furthermore, the decline in urban biodiversity is a critical consequence of urban expansion. The transformation of natural habitats into concrete jungles leads to habitat fragmentation, making it difficult for many species to survive and thrive. Urban areas are often characterized by reduced species richness and a dominance of a few generalist species that can adapt to human-modified environments [28]. This loss of biodiversity has cascading effects on ecosystem functions and services, such as pollination and pest control, which are essential for urban food production and human well-being.

In the face of these challenges, urban forestry has emerged as a crucial strategy for creating sustainable and livable cities. Urban forestry involves the planning, management, and care of trees and green spaces in urban areas to enhance the urban environment and the quality of life for city dwellers [21]. Trees in urban settings provide a range of ecosystem services that help mitigate the environmental impacts of urbanization. For instance, urban trees can significantly improve air quality by absorbing pollutants and particulate matter. A study by Nowak et al. [30] estimated that trees in the contiguous United States remove 711,000 metric tons of air pollution annually, with an associated economic value of $3.8 billion. Similarly, McDonald et al. [27] found that increasing tree cover in cities could reduce particulate pollution by up to 25%, highlighting the potential of urban forestry as a natural air purification system. In addition to air quality improvement, urban trees help regulate temperatures through the process of evapotranspiration and by providing shade. This reduces the urban heat island effect, where cities experience higher temperatures than surrounding rural areas due to heat absorption by buildings and pavement [31]. By mitigating this effect, urban trees contribute to energy savings and reduce the need for artificial cooling, thereby lowering greenhouse gas emissions.

Among the many tree species used in urban forestry, mulberry trees (genus *Morus*) stand out due to their unique combination of environmental benefits and adaptability to urban conditions [16]. Native to temperate and subtropical regions, mulberry trees are known for their rapid growth, robust nature, and ability to tolerate pollution, drought, and poor soil conditions [17]. These characteristics make them particularly well-suited for urban environments, where trees often face challenging conditions. Mulberry trees are highly efficient in capturing airborne pollutants and particulates, contributing to improved air quality. Studies have shown that mulberry leaves have a high capacity for absorbing pollutants making them effective natural air purifiers [49]. Furthermore, their dense canopy provides shade and cooling, help to reduce the urban heat island effect. In addition to their environmental benefits, mulberry trees have cultural and economic values. They have been traditionally used for silk production and their fruits are consumed for their nutritional and medicinal properties [5]. This multifunctionality enhances their appeal for urban planting, as they can contribute to food security and local economy while providing ecological services.

This chapter delves into the unique attributes of mulberry trees that make them a sustainable solution for urban forestry. It explores their ecological, environmental, and societal benefits, with a particular focus on their role in improving air quality; an issue of critical importance in many cities worldwide. It also highlights the aesthetic and functional versatility of mulberry trees. With their dense foliage and broad canopy, they provide ample shade, reducing the urban heat island effect. Furthermore, mulberry trees contribute to urban biodiversity by providing food and habitat for a variety of birds, insects, and small mammals. As cities seek sustainable solutions to enhance livability and environmental quality, mulberry trees emerge as a promising candidate for large-scale planting initiatives. Their capacity to adapt the various environmental conditions, coupled with their significant ecological benefits, positions them as a key component of future urban greening strategies.

This chapter aims to provide a comprehensive overview of the role of mulberry trees in urban forestry, discussing their potential to improve air quality, support urban biodiversity, and contribute to climate resilience. By examining case studies from different parts of the world, the chapter will reveal the successful application of mulberry trees in various urban settings. As cities continue to grapple with the dual challenges of environmental degradation and climate change, mulberry trees represent a sustainable and effective tool for creating greener, healthier, and more resilient urban environment. In the context of climate change and rapid urbanization, the integration of resilient and multifunctional tree species like mulberry into urban forestry programs is not just desirable but necessary.

2 Importance of Urban Forestry

Urban forests, or community forests, are described as a gathering of trees in cities and sub-urban areas; the urban or community forest consists of trees planted in parks, along city streets, and on residential properties. Following Covid-19, urban woodlands took on unprecedented importance, not only for leisure activities, but also for preserving social relationships and coping with psychological stress. With these various needs, many individuals have adapted urban forests in novel and unexpected ways. Botsch et al. [4] shown that woods are regarded as a "counterhorizon" to civilization, which contributes to a better understanding of recreational processes and conflict perceptions in forests. Urban forestry differs significantly from both developed and developing countries in terms of growth and development (Table 1) [22].

2.1 India in Urban Forestry

With the exception of Chandigarh and Gandhinagar, very few Indian cities have more urban forests per resident. Nevertheless, New Delhi, the Indian capital, has significant urban forestation programs [3]. Green space currently makes up 20% of Delhi's land area. The new organization in charge of urban forestry projects is the Parks and Garden Society. In Delhi, the kind of trees placed next to roads serve as markers (e.g. Vigyan path with Toona ciliata). The Gujarat state encourages tree planting by linking it to religious observances across a wide range of belief systems. Six hectares of land in Gandhinagar are covered in trees as a sign of respect for these

Table 1 Importance of urban forestry in developed and developing countries

Developed countries	Developing countries
– focuses on amenity value	– focuses on economic benefits, employment and support of agriculture in low income cities
· decreases noise	· renders food
· provides beauty and shade	· renders medicine, oil
· increases recreation value	· improves gardening conditions
· cools cities and planet	· renders fencing material
· controls water runoff	· increases cash/subsistence income
· decreases climatic extremes	· renders fodder
· improves water quality	· renders raw material, fibre
· increases health/well being	· plus all benefits of developed countries urban forestry
· provides habitat for wildlife	· renders employment
· conserves energy	· renders timber
· decreases air pollution	· renders fuel

religious convictions. It is a place set aside to install love and life in the trees for the wellbeing of the forest as a whole.

2.2 Significance of Urban Forestry

Urban forestry offers benefits such as reduced energy costs and cooler shade for houses and roadways.

- Trees help reduce pollution and improve air quality in cities by storing carbon and reducing runoff on land.
- Trees in urban environments improve home prices and earnings by providing beauty and shade.
- Providing a feeling of community and exercise opportunities can enhance the mental health and well-being of software employees in metropolitan areas, leading to reduced crime.
- Reduced energy costs of urban areas since trees provide cooling shade over properties and concrete streets.
- Pollution reduction and better air quality in cities and urban areas as a result of tree's ability for carbon storage and decreasing runoff on land.
- Increased home values, incomes, trees in landscapes definitely increase urban property values due to attractiveness and shade.
- Creating a sense of community and suitable venues for gathering and exercise can improve the mental health and welfare of software employees in urban settings, resulting in lower crime rates.
- Trees effectively reduce runoff by absorbing water through their roots. In fact, drainage from an impermeable surface is five times greater than that from a wooded watershed of equal size. Reduced runoff improves water quality because it minimizes the quantity of contaminated runoff that reaches our waterways. Urban woods help to prevent flooding and restore groundwater.
- A mature tree can absorb 48 pounds of carbon dioxide per year, which is roughly equivalent to driving 50 km. This carbon is stored in the tree's tissues until it is burned or decomposed.

Urban forestation is worthwhile because, unlike other types of construction (such as roadways, storm sewers, and electricity lines), trees gain value and quality as they grow in size rather than losing it. One of the guiding principles of urban forestry is to plant the right tree in the right location. Yellow poplar (*Liriodendron tulipifera*) is intolerant of poor urban soils, hence it is rarely used as a street tree. There is no tree species that is perfectly suited to every site, thus each species' attributes are assessed to determine their eligibility for planting as street trees. Street tree species exhibit important characteristics such as tolerance for alkaline soils, compacted soils, low soil volume, de-icing salts, drought, and good structure. Blackgum (*Nyssa sylvatica*) and swamp white oak (*Quercus bicolour*) are noted for their urban tolerance, however Blackgum is hard to transplant.

3 Why Mulberry Trees?

Urbanization has led to a rapid increase in pollution levels and environmental degradation, creating a pressing need for sustainable urban forestry solutions. Integration of mulberry trees (*Morus spp.*) into urban landscapes is one of the important solution towards sustainability [25]. Mulberry trees, known for its adaptability, resilience, and air purification properties, offer significant environmental benefits [34].

3.1 Mulberry Trees: Characteristics and Adaptability

Mulberry trees belong to the family Moraceae and include species such as *Morus alba* (white mulberry), *Morus nigra* (black mulberry), and *Morus rubra* (red mulberry). These deciduous trees are native to Asia but have become naturalized in many parts of the world. Their adaptability to various climates and soil conditions makes them an ideal tree for urban forestry [19]. Urban forestry includes managing trees and other natural systems in developed regions (e.g., cities, towns, and suburbs) for human health and well-being. Among the various tree species utilized in urban forestry, mulberry (*Morus* spp.) stands out for its adaptability, rapid growth, and environmental contributions. Beyond its well-known role in silkworm rearing and soil conservation, mulberry trees have a substantial impact on improving air quality within urban forestry systems. Followings are the mechanisms through which mulberry trees enhance air quality (Fig. 1) [32].

3.2 Carbon Sequestration and Reduction of Greenhouse Gases

One of the most critical environmental functions of mulberry trees in urban forestry is their ability to sequester atmospheric carbon dioxide (CO_2). Urban forestry systems, especially when integrated with fast-growing species like mulberry, act as carbon sinks, helping mitigate climate change [50]. The CO_2 absorbed by mulberry trees during photosynthesis is stored as biomass in the form of wood, leaves, and roots. This process reduces the concentration of greenhouse gases (GHGs) in the atmosphere, contributing to better air quality.

3.2.1 Carbon Sequestration Potential

A mature mulberry tree can absorb approximately 48 pounds of CO_2 annually, depending on the species and environmental conditions. In urban forestry systems

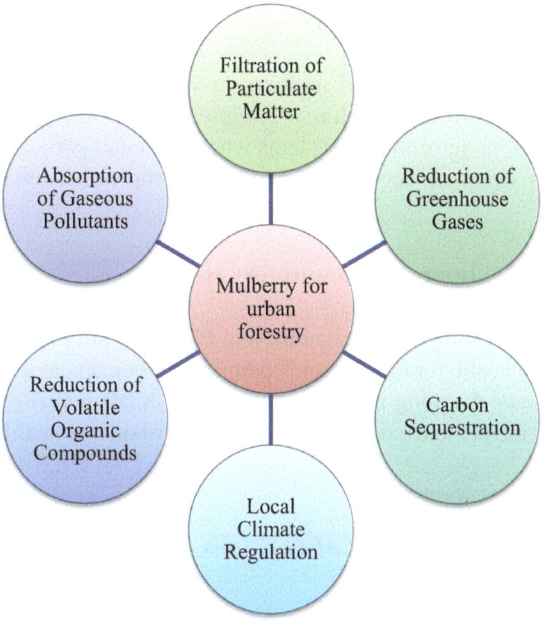

Fig. 1 Role of mulberry in improving air quality

where mulberry trees are planted alongside other plants, the cumulative effect of multiple trees enhances the system's overall carbon sequestration capacity.

3.2.2 Reduction of Methane and Nitrous Oxide

Mulberry-based urban forestry systems can also reduce the emissions of methane (CH_4) and nitrous oxide (N_2O), two potent greenhouse gases. This is achieved by enhancing soil health and minimizing the need for chemical fertilizers, which are a significant source of N_2O emissions in agriculture [51].

3.3 Filtration of Particulate Matter (PM)

Mulberry trees play a vital role in filtering particulate matter (PM), such as PM 2.5 and PM 10, which are common pollutants in rural agricultural settings. Dust, soil particles, and emissions from farming machinery contribute to the particulate load in the air. Mulberry trees, with their broad, rough-textured leaves, act as natural air filters, trapping airborne particles on the leaf surface [40].

3.3.1 Canopy Density

The dense foliage of mulberry trees creates an effective barrier that captures fine particulate matter before it can spread to surrounding areas. This is particularly beneficial in agroforestry systems where dust and particulate pollution from agricultural activities can negatively impact both crop yields and human health.

3.3.2 Improved Respiratory Health

By reducing the concentration of PM in the air, mulberry trees help improve respiratory health for farmers and communities living near urban forestry areas. Studies have shown that long-term exposure to particulate matter can lead to respiratory diseases, so the presence of mulberry trees contributes to healthier rural environments.

3.4 Absorption of Gaseous Pollutants

In urban forestry systems, where farming activities often involve the use of fertilizers and machinery, gaseous pollutants such as nitrogen oxides (NOx), sulfur dioxide (SO_2), and ammonia (NH_3) are frequently released into the atmosphere [24]. Mulberry trees absorb these harmful gases through their stomata, reducing their concentration in the air.

3.4.1 Nitrogen Oxide and Sulfur Dioxide Absorption

Mulberry leaves act as sinks for NOx and SO_2, both of which are harmful to human health and contribute to acid rain. Through the process of photosynthesis, mulberry trees absorb these gases, transforming them into less harmful compounds.

3.4.2 Ammonia Absorption

In intensive farming systems where the use of nitrogen-based fertilizers is common, ammonia emissions can contribute to air pollution. Mulberry trees, when planted in agroforestry systems, can absorb NH_3, reducing its concentration and minimizing its harmful effects on air quality.

3.5 *Reduction of Volatile Organic Compounds (VOCs)*

Volatile organic compounds (VOCs) are emitted by agricultural machinery, pesticides, and fertilizers. These compounds contribute to ground-level ozone formation, which is a significant pollutant that negatively impacts both crops and human health. Mulberry trees, through their natural biological processes, can reduce the concentration of VOCs in the air [46].

3.5.1 Buffering Effects

In an urban area, the presence of mulberry trees creates a natural buffer zone that limits the spread of VOCs from agricultural and other activities. This helps protect both the crops and the farmers working within these systems from the harmful effects of ozone and VOC exposure.

3.6 *Contribution to Local Climate Regulation*

Mulberry trees, through their shading and cooling effects, help regulate local microclimates in urban forestry systems. This regulation of temperature not only improves crop productivity but also reduces the formation of ground-level ozone, which forms at higher temperatures and is harmful to air quality [37].

3.6.1 Evapotranspiration and Cooling

Mulberry trees contribute to evapotranspiration, a process where water is absorbed from the soil and released into the atmosphere via leaf stomata. This process cools the surrounding air, reducing heat stress on trees and lowering the likelihood of harmful ozone formation.

3.6.2 Mitigating the Heat-Island Effect in Rural Areas

Although the heat-island effect is more commonly associated with urban areas, this planting system that include mulberry trees can help reduce localized heat buildup, which contributes to poor air quality. By cooling the surrounding environment, mulberry trees help maintain air quality at acceptable levels.

3.7 Support for Sustainable Farming Practices

Mulberry-based agroforestry systems are often associated with sustainable farming practices that reduce the environmental footprint of agricultural activities. For instance, mulberry leaves are used in sericulture (silkworm rearing), reducing the need for chemical inputs. Additionally, agroforestry systems that integrate mulberry trees often incorporate organic farming techniques that further reduce the emission of pollutants from synthetic fertilizers and pesticides.

3.7.1 Organic Fertilization

Mulberry trees enhance soil fertility through the cycling of organic matter, reducing the need for synthetic fertilizers that contribute to air pollution. This, in turn, minimizes the release of nitrous oxide, a potent greenhouse gas and air pollutant.

4 Case Studies: Mulberry in Urban Forestry

Mulberry trees (genus *Morus*) have been increasingly recognized as a valuable addition to urban forestry due to their diverse environmental benefits and adaptability to urban conditions. Numerous case studies from around the world have demonstrated the successful integration of mulberry trees in urban landscapes, showcasing their potential to enhance urban greenery, improve air quality, and contribute to the overall ecological and social well-being of cities. This section explores some notable case studies that highlight the role of mulberry trees in urban forestry and their impact on urban environments.

4.1 Mulberry Trees in Beijing, China: Air Pollution Mitigation [49]

Beijing, one of the most densely populated and polluted cities globally, has been grappling with severe air quality challenges for years. A study by Wu et al. [49] assessed the effectiveness of different tree species in reducing urban air pollution in Beijing and identified mulberry trees (*Morus alba)* as particularly effective. The study highlighted that mulberry leaves have a remarkable capacity to capture particulate matter (PM10 and PM2.5) and gaseous pollutants such as sulfur dioxide (SO_2) and nitrogen dioxide (NO_2). This efficacy is attributed to the dense foliage and broad leaf surface area of mulberry trees, which enable them to intercept and absorb a significant amount of airborne pollutants. Consequently, mulberry trees have become a vital component of Beijing's green infrastructure strategy to combat air pollution.

In addition to their air purification capabilities, mulberry trees have enhanced the aesthetic appeal and ecological diversity of urban green spaces in Beijing. The city's urban forestry initiatives have strategically integrated mulberry trees into various settings, including parks, streetscapes, and residential areas, showcasing their adaptability to diverse urban environments [49]. The successful use of mulberry trees in Beijing underscores their potential as a sustainable solution for improving air quality in other densely populated and polluted cities. Their versatility, resilience, and ability to thrive in harsh urban conditions make them a valuable addition to urban forestry programs worldwide. As cities continue to seek effective measures to address environmental challenges, the example of mulberry trees in Beijing provides a compelling case for their broader application in enhancing urban green spaces and mitigating air pollution.

4.2 *Mulberry as a Part of Urban Green Belt Development in Delhi, India [39]*

Delhi, the capital of India, faces extreme air pollution and a severe lack of green spaces, making urban forestry a critical strategy for environmental management. The Delhi Development Authority (DDA) initiated the development of urban green belts to mitigate the urban heat island effect, reduce air pollution, and provide recreational spaces for residents. Mulberry trees (*Morus alba* and *Morus nigra*) were strategically planted as part of this initiative due to their rapid growth, resistance to harsh urban conditions, and ability to thrive in the region's semi-arid climate. A study by Sharma et al. [39] assessed the impact of mulberry trees on air quality and microclimate in these green belts. The results showed that areas with dense mulberry plantations had significantly lower levels of ambient air pollutants, particularly $PM_{2.5}$ and NO_2, compared to surrounding areas with sparse vegetation [39]. Furthermore, the cooling effect provided by mulberry trees reduced local temperatures by up to 3 °C, making these green belts more comfortable for residents during the hot summer months. This case study underscores the importance of selecting resilient tree species like mulberry for large-scale urban greening projects in regions facing similar environmental challenges. The successful establishment of mulberry green belts in Delhi serves as a model for other cities in India and globally, where air pollution and urban heat are major concerns.

4.3 *Integrating Mulberry Trees in Urban Parks: The Istanbul Experience [6]*

Istanbul, Turkey, has a rich tradition of using mulberry trees in its urban parks and gardens, reflecting their historical and cultural significance in the region. A study conducted by Ercisli et al. [6] explored the role of mulberry trees in enhancing the biodiversity and recreational value of urban parks in Istanbul. The researchers found that mulberry trees contributed significantly to the structural diversity of urban green spaces, providing habitat and food for various bird and insect species. In addition to their ecological benefits, mulberry trees in Istanbul's urban parks have enhanced the aesthetic value and historical character of these spaces. Their attractive foliage, seasonal fruiting, and cultural associations with traditional mulberry cultivation and silk production have made them a popular choice for public spaces [6]. The study highlighted the multifunctional role of mulberry trees in urban parks, where they serve not only as shade providers and air purifiers but also as cultural symbols and educational resources. This case study illustrates the potential of mulberry trees to serve multiple functions in urban forestry beyond environmental benefits. Their integration into urban parks can enrich the cultural and recreational experience of residents, supporting both ecological and social aspects of sustainable urban development.

4.4 Urban Roadside Plantations: Mulberry Trees in Tehran, Iran [15]

Tehran, the capital city of Iran, has implemented extensive roadside plantation programs to combat air pollution and improve the urban environment. A study by Jafarian et al. [15] evaluated the effectiveness of different tree species used in these roadside plantations and found that mulberry trees (*Morus alba*) were among the most effective in reducing air pollution levels along major roadways. The study demonstrated that mulberry trees, with their dense foliage and high leaf surface area, were able to capture significant amounts of dust, soot, and other particulate matter emitted by vehicular traffic [15]. The researchers recommended the use of mulberry trees as a cost-effective and low-maintenance solution for improving air quality in urban areas with high traffic volumes. The case study also highlighted the importance of selecting appropriate species for specific urban conditions, noting that mulberry trees were well-suited to Tehran's climate and pollution levels. The successful use of mulberry trees in Tehran's roadside plantations provides valuable insights for other cities seeking to mitigate traffic-related air pollution through urban forestry. It also underscores the need for careful planning and species selection to ensure the effectiveness and sustainability of urban greening initiatives.

4.5 *Mulberry Trees in Urban Agriculture and Community Gardens: A Case from Los Angeles, USA [26]*

In Los Angeles, a city known for its sprawling urban landscape and food insecurity issues, mulberry trees have been integrated into urban agriculture and community garden projects to provide both environmental and social benefits. A study by McClintock and Cooper [26] examined the role of fruit trees, including mulberries, in community gardens across the city. The researchers found that mulberry trees were highly valued for their low maintenance requirements, high fruit yield, and resilience to urban stressors such as poor soil quality and limited water availability. Mulberry trees in these community gardens contributed to local food security by providing fresh fruit to residents, many of whom had limited access to affordable, healthy food [26]. The trees also supported community engagement and environmental education, as residents participated in planting, caring for, and harvesting the trees. The study highlighted the multifunctional benefits of mulberry trees in urban agriculture settings, where they serve as a source of nutrition, a tool for community building, and a component of the urban forest [26]. This case study from Los Angeles demonstrates the potential of mulberry trees to contribute to urban food systems and community well-being, in addition to their traditional roles in urban forestry. Their integration into community gardens and urban farms offers a model for other cities looking to enhance food security and community resilience through urban greening.

4.6 *Recycling Urban Waste: Mulberry Cultivation on Sludge-Composted Soils [42]*

The integration of urban sludge composted with garden waste as a soil amendment presents an innovative method for sustainable urban sludge disposal, and mulberry (*Morus alba* L.) has shown promising suitability for growth in such amended soils. A study by Si et al. [42] evaluated the effectiveness of using composted urban sludge mixed with garden waste for cultivating mulberry, demonstrating significant benefits for both soil quality and plant performance. The research highlighted that urban sludge, often rich in organic matter and nutrients, when composted with garden waste, becomes a valuable soil amendment. The study found that this compost significantly improved the soil's physical properties, such as porosity and water-holding capacity, while also enhancing its chemical attributes, including increased levels of essential nutrients like nitrogen, phosphorus, and potassium. These improvements create an ideal growth medium for mulberry, which thrives in nutrient-rich, well-drained soils. The study by Si et al. [42] showed that mulberry trees grown on soil amended with the composted sludge-garden waste mixture exhibited robust growth, higher leaf biomass, and improved chlorophyll content compared to those grown on untreated soils. This indicates that the amended soils not only support healthy mulberry growth but also enhance the plant's physiological performance. Furthermore, the mulberry trees were able to tolerate and even accumulate some heavy metals present in the sludge, making them a viable option for phytoremediation.

The findings suggest that using composted urban sludge with garden waste for mulberry cultivation can serve as a sustainable waste management strategy, converting waste into a resource while supporting urban greening initiatives. This approach not only mitigates the environmental impact of sludge disposal but also contributes to the sustainable production of mulberry for various agricultural and industrial uses. The research provides a foundation for further exploration into the use of such composts in urban and peri-urban agriculture.

These case studies illustrate the diverse applications and benefits of mulberry trees in urban forestry across different global contexts. From air pollution mitigation and temperature regulation to supporting urban biodiversity and food security, mulberry trees have proven to be a versatile and resilient addition to urban landscapes [1]. Their successful integration into various urban settings demonstrates their potential to contribute to sustainable urban development and improve the quality of life for city dwellers. As cities continue to grapple with the environmental and social challenges of rapid urbanization, mulberry trees offer a promising solution for creating greener, healthier, and more resilient urban environments. Future research and urban forestry initiatives should continue to explore the potential of mulberry trees and other multifunctional species to address the complex needs of urban ecosystems and communities.

5 Mulberry Trees and Climate Adaptation

Mulberry (*Morus* spp.) is a fast-growing, deciduous perennial plant from the *Moraceae* family and is widely acknowledged for its significant economic role in producing the silk through feeding of its leaves to silkworm (*Bombyx mori*) larvae [36]. It is a versatile plant that thrives in various climates and soil conditions exhibiting a remarkable adaptability that allows it to flourish in both wild and cultivated forms across diverse geographical regions. Its distribution ranges from sea level to altitudes of 4000 m above mean sea level [5, 20, 38]. This extensive adaptability of mulberry is indicative to varied climatic conditions enabling its cultivation in mountainous, plain and valley terrains under both rain-fed and irrigated conditions [44]. Furthermore, mulberry can endure harsh environments, including humid and semi-arid lands and can be cultivated in various forms such as bush, dwarf and tree.

Its resilience to low water requirements and tolerance to pollution makes it an ideal tree for urban settings including home gardens, city landscaping and roadside planting for shade and aesthetic enhancement. Historically, mulberry has been recognized primarily for its role in silk production and as feed for domesticated animals, however, its ecological significance has often been overlooked. Mulberry plays a crucial role in environmental protection through the eco-restoration of degraded lands, bioremediation of polluted sites and air purification via carbon sequestration. Its dense and deeply rooted network contributes to soil and water conservation [13,

33]. The drought-resistant characteristics of mulberry make it suitable for use as avenue trees, windbreaks and biomass belts where it aids in reducing wind speed and providing organic material. Furthermore, mulberry leaves possess a remarkable capacity to absorb pollutants such as chloride, hydrogen fluoride and sulphur dioxide thereby enhancing air quality in polluted environments. Research indicates that mulberry leaves remain undamaged even in high atmospheric chloride levels [23].

In recent years, mulberry has gained recognition as a multipurpose plant due to its contributions to environmental safety, medicinal properties and industrial applications across various sectors. The biologically active compounds found in the leaves, stems and roots of mulberry enhance human health and well-being [2, 45]. The diverse applications of mulberry extend to the pharmaceutical, food, cosmetic and healthcare industries where every part of the plant is utilized in the production of various products [52, 53].

Climate Adaptations

Mulberry trees exhibit a variety of adaptations that enhance their resilience to changing climatic conditions. Their robust growth and adaptability in diverse environments make them a vital resource in areas affected by climate variability. Key climate adaptations of mulberry trees, including drought tolerance, temperature resilience, and pollution resistance.

5.1 Drought Tolerance

A prominent adaptation of mulberry trees is their ability to endure drought conditions. They develop deep taproots that enable access to water from deeper soil layers, enhancing their resilience during water scarcity. This trait is particularly advantageous in arid and semi-arid regions where water availability is inconsistent. Once established, mulberry trees can withstand prolonged dry periods, making them suitable for cultivation in areas increasingly affected by drought due to climate change [10].

5.2 Tolerance to Extreme Temperatures

Mulberry trees are capable of thriving in a wide range of temperatures, flourishing in both temperate and subtropical climates. Optimal growth occurs at temperatures between 24 and 28 °C, but they can tolerate fluctuations from 13 upto 39 °C. Their resilience to temperature variations allows them to adapt to the unpredictable patterns associated with global warming. However, extreme temperatures can impact growth, necessitating careful selection of mulberry varieties suited to specific climatic conditions [10].

5.3 Resilience to Air Pollution

The leaves of mulberry trees have a unique capacity to absorb specific air pollutants, including chlorine, hydrogen fluoride, and sulfur dioxide. This characteristic makes them well-suited for urban and industrial environments where air pollution is prevalent. Research indicates that mulberry leaves can remain undamaged even under elevated chlorine levels, allowing them to thrive in polluted areas and function as natural air purifiers, thereby improving local air quality [9, 23].

5.4 Flood Tolerance

While mulberries are drought-resistant, they also exhibit moderate resilience to short-term flooding. Their robust root systems enable them to cope with temporary waterlogging without significant damage. This trait is particularly beneficial in regions prone to sudden heavy rains or flooding events linked to climate change. Although prolonged water logging can be detrimental, their ability to withstand brief flooding enhances their adaptability in increasingly unpredictable climates [12, 29].

5.5 Wind Resistance

Mulberry trees possess a sturdy structure that provides strong resistance to wind, making them suitable for planting in areas prone to high winds or storms. Their wind tolerance allows them to serve as effective windbreaks, reducing wind speeds, protecting crops, and preventing soil erosion. This characteristic is especially valuable in regions where climate change has led to an increase in extreme weather events, such as cyclones and storms [38].

5.6 Adaptability to Soil Conditions

Mulberry trees demonstrate high adaptability to various soil types, including well-drained loamy, sandy, and slightly acidic soils. Their ability to thrive in different soil conditions, even those of marginal quality, contributes to soil health by enhancing fertility through litter fall, which enriches the soil with organic matter. This adaptability is crucial as climate change alters soil moisture levels, salinity, and fertility in many regions [10].

5.7 *Efficient Water Use*

Once established, mulberry trees have low water requirements, allowing them to grow with minimal irrigation in water-scarce regions. Their efficient use of available water makes them particularly suitable for areas with erratic rainfall patterns, where water conservation is essential. Techniques such as mulching and drip irrigation can further enhance their water-use efficiency, enabling them to cope with reduced rainfall due to climate change [10].

5.8 *Pruning Tolerance and Resilience*

Mulberry trees exhibit a high tolerance for pruning, allowing them to be regularly cut back without detrimental effects on overall growth. This trait is beneficial in regions where farmers employ "chop and drop" techniques to enrich soil or where space for tree growth is limited. Their ability to recover quickly from pruning supports adaptation to changing agricultural practices aimed at mitigating climate impacts [38].

5.9 *Quick Growth for Reforestation and Biomass Production*

The rapid growth of mulberry trees makes them suitable for reforestation initiatives in degraded landscapes and for establishing biomass belts. Their swift growth facilitates the quick establishment of vegetation cover, which can mitigate climate change effects by enhancing carbon sequestration, reducing soil erosion, and improving local microclimates. Additionally, mulberry's biomass can serve as a valuable resource for renewable energy in regions transitioning to sustainable energy practices [38].

5.10 *Salt Tolerance*

Mulberry trees exhibit moderate salt tolerance, enabling them to thrive in saline soils, which are increasingly common due to rising groundwater salinity and salt-water intrusion in coastal areas. While they prefer well-drained loamy soils, certain mulberry varieties can tolerate slightly saline conditions without significant reductions in growth or leaf quality. This ability is crucial in regions facing increasing soil salinity due to climate change, allowing mulberries to survive where other crops may fail [38].

5.11 Coppicing Ability

Mulberry trees possess a strong capacity for regeneration through coppicing, where the tree is cut back to ground level, and new shoots rapidly grow from the stump. This adaptation allows mulberries to recover quickly from damage caused by natural events such as storms or human activities like logging. The ability to regenerate enhances their utility in agricultural systems and reforestation efforts, ensuring long-term productivity in variable and harsh climatic conditions [38].

5.12 Allelopathy

Mulberry trees exhibit allelopathic properties, releasing specific biochemicals into the soil that inhibit the growth of surrounding plants. This adaptation reduces competition for resources such as water, nutrients, and sunlight. In regions experiencing climate stress, where competition for limited resources is heightened, allelopathy provides mulberries with a competitive advantage. By suppressing the growth of weeds or invasive species, mulberry trees can sustain their growth in degraded or overpopulated environments, enhancing their resilience in agriculture and ecological restoration efforts [11].

6 Challenges and Considerations of Urban Forestry

6.1 Constraints of Urban Forestry

1. Urban trees have challenges such as pollution, high temperatures, drought, and floods. Additionally, their restricted root area makes them vulnerable to insects and diseases.
2. To maintain urban woods effectively, numerous stakeholders and municipal governments must collaborate due to limited resources.
3. Urban growth often leads to the loss of green places and forests.
4. Promoting urban woods might be challenging due to a lack of public knowledge of their advantages. In example, urban trees are frequently regarded as a financial burden or risk.
5. Mulberry trees offer numerous benefits, but on the other side their use in urban forestry also comes with some challenges as mentioned below:

   5.1 ***Invasive species***: The paper mulberry tree, which was introduced to North America in the early twentieth century, is considered invasive in the US. It has a fast growth rate and shallow root system, which can make it susceptible to structural issues.

5.2 ***Pests***: The red mulberry tree is susceptible to a variety of pests, including spider mites, powdery mildew, leaf spots, canker diseases, and scale insects.
5.3 ***Fruit Drop and Mess***: Some species produce large amounts of fruit, which can create a mess on streets and sidewalks. However, fruitless cultivars of mulberry are available and widely used in urban settings.
5.4 ***Allergy Concerns***: The pollen of some mulberry species can cause allergies in sensitive individuals. Selecting low-allergen varieties can mitigate this issue.
5.5 ***Maintenance***: Regular pruning and maintenance are necessary to manage their rapid growth and ensure they do not interfere with infrastructure.

6.2 Future Attentions

Shear [41] identifies three areas for future study based on developments in remote sensing, data collecting, and processing.

1. ***The relationship between LiDAR and urban–rural forestry*** analysis and benefit evaluation tools: The majority of urban forestry management software available today maps forest features directly from high-resolution images. It will be desirable to combine LiDAR or PhoDAR with high-resolution multispectral photography to get structural properties of urban forests for assessing ecosystem services.
2. ***Real-Time Remote Sensing***: Time-sensitive and dynamic phenomena, such as flash flooding, storms, and forest disease, frequently endanger urban forest ecosystems. To protect human and ecological resources, these occurrences must be monitored continuously, visualised, and analysed quickly. The typical technique involves categorising remotely detected pictures. However, employing real-time remote sensing data as input can aid in the prediction, visualisation, and identification of hot regions of occurrence [35].
3. ***Mulberry trees can aid in afforestation*** efforts by improving soil conditions, conserving water, and improving air quality.
4. ***Roadside Planting***: Mulberry trees can be planted along roadsides, in parks, and in gardens.
5. ***Intercropping***: Mulberry trees can be intercropped with other crops, such as vegetables or grass.
6. ***Flood Protection***: Mulberry plantations can help prevent soil erosion during floods and water logging conditions.
7. ***Sand Storm Suppression***: Mulberry plantations can help suppress sand storms.

Finally, as research expands the utility of remote sensing to support urban forestry with mulberry, a stronger connection between urban foresters, remote sensing experts, and programmers is critical to ensuring optimal provision of ecosystem services for well-being by urban forests.

7 Future Prospects and Recommendations

Mulberry trees (*Morus* spp.) have demonstrated significant potential as a sustainable solution for enhancing urban forestry and improving air quality. To fully harness their benefits and ensure their successful integration into urban environments, several strategies and recommendations should be considered.

7.1 Selection of Suitable Cultivars

The choice of appropriate mulberry cultivars is critical for maximizing their benefits in urban settings. While *Morus alba* is widely known for its adaptability, selecting cultivars that are fruitless or have reduced fruiting potential can mitigate issues related to fruit drop and mess. Additionally, cultivars with low allergenic potential and high tolerance to pollution and varying climatic conditions should be prioritized. Research and development of new cultivars specifically suited to urban conditions could further enhance the role of mulberry trees in city landscapes.

7.2 Community Engagement and Education

The success of urban forestry initiatives depends significantly on public support and participation. Educating communities about the ecological and health benefits of mulberry trees, such as their role in reducing air pollutants and providing shade, can foster a sense of ownership and responsibility. Involving community members in planting, care, and maintenance activities can strengthen community ties and ensure the long-term sustainability of these initiatives. Programs that engage schools, local organizations, and volunteer groups can be particularly effective.

7.3 Integration into Urban Planning

For mulberry trees to be truly effective in urban forestry, they must be incorporated into comprehensive urban planning strategies. This includes careful site selection, considering factors like air quality improvement potential, soil conditions, and available space. Urban planners should integrate mulberry trees into green corridors, parks, and streetscapes, ensuring that they complement existing infrastructure and contribute to broader environmental goals. Furthermore, developing guidelines for the maintenance and management of mulberry trees can help municipalities and residents maximize their benefits while minimizing potential challenges.

7.4 Research and Monitoring

Continuous research on the performance of mulberry trees in urban settings is essential for refining best practices and making data-driven decisions. Monitoring growth, pollutant absorption efficiency, and ecological impact can provide valuable insights into the long-term viability and benefits of mulberry trees in various urban environments. Collaboration between academic institutions, government agencies, and non-profit organizations can facilitate the sharing of knowledge and resources.

7.5 Policy Support and Incentives

Local governments should consider policies that support the planting and maintenance of mulberry trees in urban areas. Incentives for property owners and developers to include mulberry trees in their projects, as well as funding for public planting initiatives, can significantly boost the adoption of these trees in cities. Furthermore, integrating urban forestry into climate action and air quality management plans can align the use of mulberry trees with broader environmental and public health goals. By implementing all these strategies, cities can effectively leverage the unique benefits of mulberry trees, contributing to healthier, greener, and more resilient urban environments.

8 Conclusion

As urban areas continue to expand and environmental challenges intensify, integrating resilient and multifunctional tree species like mulberry into urban forestry programs is not just desirable but necessary. Mulberry trees, with their capacity to improve air quality, support urban biodiversity, and adapt to harsh conditions, offer a promising solution for creating greener, healthier, and more resilient urban environments. This chapter aims to provide a comprehensive overview of the potential of mulberry trees in urban forestry, supported by research evidence and practical examples from cities around the world. By understanding the unique benefits and challenges of mulberry trees, urban planners and policymakers can make informed decisions to enhance the sustainability and livability of urban landscapes. Mulberry trees offer a sustainable and effective solution for urban forestry, contributing to improved air quality, climate resilience, and urban biodiversity. By carefully selecting and managing these trees, cities can create healthier, more livable environments for their inhabitants. As the challenges of urbanization and climate change continue to grow, the role of resilient tree species like mulberry in urban landscapes will become increasingly important.

References

1. Ali S, Khan SM, Abdullah A, Rana M, Ahmad Z (2024) Dryland agroforestry: mitigating role in reducing air pollution and climate change impacts. In: Agroforestry for carbon and ecosystem management. Academic Press, pp 271–282
2. Bao T, Xu Y, Gowd V, Zhao J, Xie J, Liang W, Chen W (2016) Systematic study on phytochemicals and antioxidant activity of some new and common mulberry cultivars in China. J Funct Foods 25:537–547. https://doi.org/10.1016/j.jff.2016.07.001
3. Bherwani H, Banerji T, Menon R (2024) Role and value of urban forests in carbon sequestration: review and assessment in Indian context. Environ Dev Sustain 26(1):603–626
4. Botsch K, Wurster M, Ensinger K, Selter A, Bethmann S (2014) Metaphorische repräsentationen des waldes. Allg Forst Jagdztg 189:187–202
5. Ercisli S, Orhan E (2007) Chemical composition of white (*Morus alba*), red (*Morus rubra*), and black (*Morus nigra*) mulberry fruits. Food Chem 103:1380–1384. https://doi.org/10.1016/j.foodchem.2006.10.054
6. Ercisli S, Orhan E, Hizarci Y, Yildirim N, Agar G (2008) Genetic diversity in grapevine germplasm resources in the Coruh valley revealed by RAPD markers. Biochem Genet 46:590–597
7. Forte B, Cerreta M, De Toro P (2019) The human sustainable city: challenges and perspectives from the habitat agenda
8. Ghorani-Azam A, Riahi-Zanjani B, Balali-Mood M (2016) Effects of air pollution on human health and practical measures for prevention in Iran. J Res Med Sci 21:65
9. Ghosh M, Mukhopadhyay A, Mukhopadhyay UK (2013) Effect of air pollution on physiology and yield of mulberry silkworm, *Bombyx mori* L. Res J Chem Environ 17(5):60–68
10. Guha A, Sengupta D, Rasineni GK, Reddy AR (2010) An integrated diagnostic approach to understand drought tolerance in mulberry (*Morus indica* L.). Flora 205(2):144–151
11. Haq RA, Hussain M, Cheema ZA, Mushtaq MN, Farooq M (2010) Mulberry leaf water extract inhibits Bermuda grass and promotes wheat growth. Weed Biol Manag 10(4):234–240
12. Hu J, Duan Y, Yang J, Gan L, Chen W, Yang J et al (2023) Transcriptome analysis reveals genes associated with flooding tolerance in mulberry plants. Life 13(5):1087
13. Huang F, Wang D (2012) The introduction to mulberry to ecological restoration. North Seric 33(4):52–54
14. Izah SC, Iyiola AO, Yarkwan B, Richard G (2023) Impact of air quality as a component of climate change on biodiversity-based ecosystem services. In: Visualization techniques for climate change with machine learning and artificial intelligence. Elsevier, pp 123–148
15. Jafarian A, Mirzaei F, Zare A (2015) Evaluating the effectiveness of different tree species in removing air pollution in Tehran. J Environ Manag 150:9–17
16. Jiang Y, Huang R, Yan X, Jia C, Jiang S, Long T (2017) Mulberry for environmental protection. Pak J Bot 49(2):781–788
17. Kafkas S, Koşar M, Tümen G, Başer KHC (2008) Analysis of volatile components from leaves, fruits, and flowers of two Turkish mulberry species (*Morus alba* L. and *Morus nigra* L.). Flavour Fragr J 23(6):392–396
18. Van Kamp I, Leidelmeijer K, Marsman G, De Hollander A (2003) Urban environmental quality and human well-being: towards a conceptual framework and demarcation of concepts; a literature study. Landsc Urban Plan 65(1–2):5–18
19. Kaushal R, Kumar A, Mandal D, Tomar JMS, Jinger D, Islam S, Madhu M (2024) Mulberry based agroforestry system and canopy management practices to combat soil erosion and enhancing carbon sequestration in degraded lands of Himalayan foothills. Environ Sustain Indic 100467
20. Khan MA, Rahman AA, Islam S, Khandokhar P, Parvin S, Islam MB et al (2013) A comparative study on the antioxidant activity of methanolic extracts from different parts of *Morus alba* L. (Moraceae). BMC Res Notes 6:24
21. Konijnendijk CC, Nilsson K, Randrup TB, Schipperijn J (eds) (2006) Urban forests and trees: a reference book. Springer Science & Business Media

22. Kuchelmeister G (1993) Trees, settlements and people in developing countries. Arboric J. 17(4):399–411
23. Lu M, Wang RQ, Qi XS (2004) Reaction of planting tree species on chlorine pollution in the atmosphere. J Shandong Univ 39:98–101
24. Ma L, Dong Y, Du X, Yang Q, Zou K (2024) Releasing characteristics and influencing factors analysis of nitrogen-containing gaseous pollutants in wastewater treatment plants with sequencing batch reactor process. J Environ Chem Eng 2:114666
25. Manzoor S, Qayoom K (2024) Environmental Importance of Mulberry: a review. J Exp Agric Int 46(8):95–105
26. McClintock N, Cooper J (2010) Cultivating the commons: an assessment of the potential for urban agriculture on public land in Los Angeles. Cities 27(4):265–277
27. McDonald AG, Bealey WJ, Fowler D, Dragosits U, Skiba U, Smith RI et al (2007) Quantifying the effect of urban tree planting on concentrations and depositions of PM10 in two UK conurbations. Atmos Environ 41(38):8455–8467
28. McKinney ML (2008) Effects of urbanization on species richness: a review of plants and animals. Urban Ecosyst 11:161–176
29. Mo R, Zhang N, Hu D, Jin Q, Li J, Dong Z et al (2022) Identification of phenological growth stages of four Morus species based on the extended BBCH-Scale and its application in fruit development with morphological profiles and color characteristics. Horticulturae 8(12):1140
30. Nowak DJ, Crane DE, Stevens JC (2006) Air pollution removal by urban trees and shrubs in the United States. Urban For Urban Green. 4(3–4):115–123
31. Oke TR (1982) The energetic basis of the urban heat island. Q J R Meteorol Soc 108(455):1–24
32. Oksanen E, Kontunen-Soppela S (2021) Plants have different strategies to defend against air pollutants. Curr Opin Environ Sci Health 19:100222
33. Qin J, He N, Wang Y, Xiang Z (2012) Ecological issues of mulberry and sustainable development. J Resour Ecol 3(4):330–339
34. Rahman MS, Islam SMS (2021) Studies on food, health and environmental perspectives in Mulberry (Morus spp.)—a review. J Bio-Sci 29:163–179
35. Rochon GL, Biehl L, Fall S, Grant D, Thiam T, Araya B et al (2010) Real-time remote sensing in support of ecosystem services and sustainability. In: Achieving environmental security: ecosystem services and human welfare. IOS Press, pp 217–23
36. Rohela GK, Shukla P, Kumar R, Chowdhury SR (2020) Mulberry (*Morus* spp.): an ideal plant for sustainable development. Trees For People 2:100011
37. Sarkar T, Doss SG, Sivaprasad V, Teotia RS (2021) Stress tolerant traits in mulberry (*Morus* spp.) resilient to climate change: an update on its genetic improvement. Mulberry, 165–179
38. Sarkar T, Mogili T, Sivaprasad V (2017) Improvement of abiotic stress adaptive traits in mulberry (Morus spp.): an update on biotechnological interventions. 3 Biotech, 7:1–14
39. Sharma P, Gupta A, Dhillon M (2018) Green belts for sustainable urban development: the case of Delhi. Environ Monit Assess 190(7):392
40. Sharma P, Yadav P, Ghosh C, Singh B (2020) Heavy metal capture from the suspended particulate matter by *Morus alba* and evidence of foliar uptake and translocation of PM associated zinc using radiotracer (65Zn). Chemosphere 254:126863
41. Shear D (2014) PhoDAR, LiDAR and forestry: using SfM for canopy height modeling-a case study. In: AGU fall meeting abstracts
42. Si L, Peng X, Zhou J (2019) Evaluation of mulberry (Morus alba L.) growth on soils amended with urban sludge composted with garden waste. J Environ Manage, 242:261–267
43. Singh P, Yadav D (2021) Link between air pollution and global climate change. In: Global climate change. Elsevier, 79–108
44. Srivastava R, Kapoor A, Thathola A, Srivastava RP (2003) Mulberry (*Morus alba*) leaves as human food: a new dimension of sericulture. Int J Food Sci Nutr 54:411–416
45. Venkatesh KR, Chauhan S (2008) Mulberry: life enhancer. J Med Plants Res. 2(10):271–278
46. Wang Q, Shen J, Zeng B, Wang H (2021) Research on VOCs and odor from heartwood and sapwood of paper mulberry (*Broussonetia papyrifera* (L.) Vent.) with different moisture content. Wood Sci Technol 55(4):1153–1170

47. Wolch JR, Byrne J, Newell JP (2014) Urban green space, public health, and environmental justice: the challenge of making cities 'just green enough.' Landsc Urban Plan 125:234–244
48. World Health Organization (2017) Inheriting a sustainable world? World Health Organization, Atlas on children's health and the environment
49. Wu Y, Zhang S, Zhang Z (2017) The effect of urban green space on air pollution and its health benefits: evidence from Beijing. J Environ Sci 62:33–43
50. Yadav VK, Padhan D, Sen S, Gayathri T, Praveen Kumar K, Tewary P (2020) Assessment of carbon sequestration potential of mulberry vis-à-vis soils under different cultivation practices. Int J Curr Microbiol Appl Sci. 9:2463–2474
51. Yang X, Duan P, Liu Q, Wang K, Li D (2024) Addition of cellulose and hemi cellulose degrading microorganisms intensified nitrous oxide emission during composting. Bioresour Technol 393:130100
52. Yang X, Yang L, Zheng H (2010) Hypolipidemic and antioxidant effects of mulberry (*Morus alba* L.) fruit in hyperlipidaemia rats. Food Chem Toxicol 48:2374–2379
53. Zhang H, Ma ZF, Luo X, Li X (2018) Effects of mulberry fruit (*Morus alba* L.) consumption on health outcomes: a mini-review. Antioxid Basel 7(5):69

The manufacturer's authorised representative in the EU is Springer Nature Customer Service Centre GmbH, Europaplatz 3, 69115 Heidelberg, Germany. If you have any concerns regarding our products, please contact ProductSafety@springernature.com

Printed and bound by CPI Group (UK) Ltd, Croydon, CR0 4YY

01/10/2026

02225923-0006